Also by Jay Ingram

The Science of Why²: Answers to Questions About the Universe, the Unknown, and Ourselves

The Science of Why: Answers to Questions About the World Around Us

The End of Memory: A Natural History of Aging and Alzheimer's

Fatal Flaws: How a Misfolded Protein Baffled Scientists and Changed the Way We Look at the Brain

Theatre of the Mind: Raising the Curtain on Consciousness

Daily Planet: The Ultimate Book of Everyday Science

The Daily Planet Book of Cool Ideas: Global Warming and What People Are Doing About It

The Science of Everyday Life

The Velocity of Honey: And More Science of Everyday Life

The Barmaid's Brain and Other Strange Tales from Science

The Burning House: Unlocking the Mysteries of the Brain

A Kid's Guide to the Brain

Talk Talk Talk: Decoding the Mysteries of Speech

It's All in Your Brain

Real Live Science: Top Scientists Present Amazing Activities Any Kid Can Do

Amazing Investigations: Twins

The Science of Why

Volume 3

Answers to Questions About Science Myths, Mysteries, and Marvels

Jay Ingram

PUBLISHED BY SIMON & SCHUSTER
NEW YORK   LONDON   TORONTO   SYDNEY   NEW DELHI

SIMON &
SCHUSTER
CANADA

Simon & Schuster Canada
A Division of Simon & Schuster, Inc.
166 King Street East, Suite 300
Toronto, Ontario M5A 1J3

This Simon & Schuster Canada edition November 2018

SIMON & SCHUSTER CANADA and colophon are registered trademarks
of Simon & Schuster, Inc.

For information about special discounts for bulk purchases, please contact Simon & Schuster
Special Sales at 1-800-268-3216 or CustomerService@simonandschuster.ca.

Illustrations by Tony Hanyk, tonyhanyk.com
Interior design by David Gee

Manufactured in the United States of America

10 9 8 7 6 5 4 3 2 1

Library and Archives Canada Cataloguing in Publication

Ingram, Jay, author
The science of why, Volume 3 : answers to questions about science myths, mysteries, and marvels / Jay Ingram.

Issued in print and electronic formats.
ISBN 978-1-5082-5795-0 (hardcover).—ISBN 978-1-5082-5796-7 (ebook)
                    1. Science—Popular works. 2. Science—Miscellanea. I. Title.
Q162.I553 2018   500   C2018-902297-3
                                C2018-902298-1

ISBN 978-1-5082-5795-0
ISBN 978-1-5082-5796-7 (ebook)

This book is for Oscar Alexander Quinlan, even though I doubt he'll read it for a few years

Contents

Part 1: Amazing Animals

Part 2: The Body

Part 3: Weird, Wacky, and Wondrous

Part 4: Mystery Meals

The Science of Why
Volume 3

Part 1
Amazing Animals

Why do birds stand on one leg?

THIS QUESTION REALLY HAS TWO PARTS: WHY AND HOW? Even if you can come up with a good reason that a two-legged creature would spend time on one leg, there's still the issue of how it could do that for lengthy periods of time. If you don't think that's difficult, try standing on one leg, sticking your arms out sideways, then closing your eyes. Not easy!

Not all birds stand on one leg—mostly it's the long-legged ones, like cranes, storks, and especially flamingos. More often than not, because these birds spend a fair amount of time feeding in oceans, lakes, and rivers, the standing leg is submerged in water. And this has led scientists to think about heat loss.

Birds use a trick of anatomy to minimize the loss of body heat through their feet and legs.

Sometimes I just need a leg up.

The arteries carrying warm blood to their feet are adjacent to veins bringing blood that has been cooled by exposure to the cold air—or in this case, water—back into the body. The warm and cool blood exchange heat, ensuring that the blood entering the bird's body is warmer than it would have been, thus reducing the amount of heat loss and minimizing the energy the bird has to put into maintaining its body temperature.

You don't have to put a bird in the lab to see this: ducks and gulls do not leave footprints when they stand on ice or snow. Their feet are too cool to do that.

It's not a bad system, but some studies have shown that heat loss in the water is four times greater than in the air, so perhaps standing on one leg would be a help. Instead of two legs returning cool blood to the body, there would be only one, with the other tucked in under the body to keep it warm.

With that thinking in mind, Matthew Anderson at St. Joseph's University in Philadelphia observed captive flamingos at the Philadelphia Zoo to get a better picture of what's going on. He and his colleagues found that flamingos were much more likely to stand on one leg if they were standing in water rather than on land. Anderson interpreted this to mean that the birds were doing so to reduce cooling. And further experiments showed that this was true: the lower the temperature, the more time the birds spent on one leg.

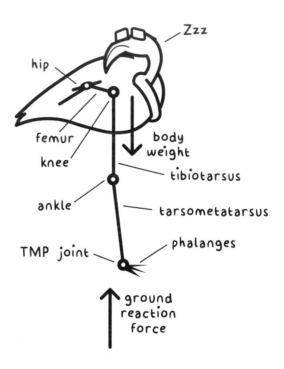

Diagram of a Sleeping Flamingo

Zzz

hip

femur

knee

body weight

tibiotarsus

ankle

tarsometatarsus

phalanges

TMP joint

ground reaction force

If it's true that flamingos stand on one leg to minimize heat loss, then when the weather is very warm you might expect they'd prefer to stand on both legs to cool down. That's exactly what Anderson found. Anderson and his colleagues also found that flamingos exhibited no preference for one foot over the other, even though other birds, like curlews and avocets, seem to prefer standing on their right foot.

So the prevailing opinion right now is that birds like flamingos stand on one leg to minimize heat loss. But isn't it difficult to balance like this? In a very cool study at the Georgia Institute of Technology, Young-Hui Chang and Lena Ting showed that the flamingo is beautifully adapted to stand on one leg, without any effort!

If this debate about why I tuck away one leg goes on any longer, I'm going to have to put my foot down!

They started their investigation at least partly because there's a potential drawback to standing on one leg to stay warmer. If it requires constant adjusting of muscles to maintain balance, the energy expenditure might be too much to justify the one-legged stance.

Ting and Chang experimented both with the bodies of two dead flamingos and with a selection of live birds to identify the demands of one-leggedness. This is easier to understand when you know that birds like flamingos actually stand on their toes. What looks like a backward-bending knee is really the ankle, and the knee joint is usually hidden by the feathers of the body. Higher up, the thighbone, unlike ours, is positioned almost horizontally. So when they are standing, they're sort of crouching.

Even a dead flamingo can do this. The researchers found that when the bird was positioned on one leg, with its center of gravity just in front of the knee (which is tucked up into the body), the joints, especially the knee and hip, lock into place, and can support the weight of the bird easily. A living bird would have to exert very little, if any, effort to stay stable in this position.

Chang and Ting met a couple of surprises along the way. First, they were sure that they knew which muscle groups were engaged when the birds stood on one leg. So they set up a high-speed camera to witness the collapse of the dead bird when they severed the muscle. But the bird didn't budge! Then when Chang hoisted the dead flamingo into the air by the shin, the leg snapped into its rigid position like a tent pole.

With living birds, this locking mechanism engages only when the bird adjusts its body into position over one leg. And it allows birds to rest on the locked leg with almost no wobbling around the center of balance, no more than a centimeter. In fact, flamingos appear to expend very little energy maintaining their one-legged stance, an observation that moved Chang and Ting to make a radical suggestion: the birds don't stand on one leg to reduce heat loss, but to reduce muscular activity, the energy-burning sort that would help maintain balance if the birds were standing on both legs. They can't be sure this is true, but this alternative explanation might help explain why a wide variety of birds, including many who live in the tropics and therefore shouldn't have to worry about cooling off too much, nevertheless still stand on one leg.

Can sharks really smell a drop of blood from a mile away?

No, THEY CAN'T.

Before the disappointment sets in, let me say that they do have amazing abilities. With large areas of nasal and brain tissue devoted to detecting odors, sharks are indeed able to detect tiny numbers of molecules in huge volumes of water. It's just that whoever said they can smell a drop of blood a mile away set them up for failure. But let's take a look at how close they might get.

Why is everyone out for blood?

Every animal is largely built of proteins, so when studying a shark's ability to sense prey in the ocean, scientists look at its sensitivity to amino acids, the building blocks of proteins. The amino acids in a drop of blood are largely those of hemoglobin, the protein that carries oxygen all over the body and helps return carbon dioxide to the lungs where it is exhaled. There are 300 million hemoglobin molecules in each red blood cell, and 5 million red blood cells in each drop of blood. Every hemoglobin molecule contains 546 amino acids. Multiply all that together and a drop of blood contains about 800,000,000,000,000,000 (800 quintillion) amino acids. That's a lot. But it's a big ocean.

Let's put our shark a mile away from a drop of blood on the ocean surface. Currents will determine how quickly and in what direction the drop will be diluted, but for the sake of argument, let's say that the ocean is fairly calm and the amino acids spread out in all directions: north, south, east, west, and down, slowly filling a half-sphere of ocean water that has a radius of a mile (1.6 kilometers). A gargantuan bowl of ocean water. By the time any of the amino acids have drifted into the shark's range, they will have been dramatically diluted. That half-sphere of water would be about 8.5 cubic kilometers, or 8.5 trillion liters. So now, on average, there will be less than 100,000 amino acids for every liter of ocean water. (And of course this is just the math; there would be no way of predicting exactly how such a drop of blood would spread.)

That's the challenge for the shark. Are they up to it?

It doesn't look like it. Several experiments have been done with different species of sharks, and while these are not anything like the experience in the open ocean, they at least set some boundaries on what the animal could do.

A typical experiment is to immobilize a shark in water, then record electrical activity from the olfactory apparatus in its brain. If the signal spikes, they've detected the amino acids that are being added to the flow of water passing through the shark's nostrils. The results suggest that sharks are very sensitive to these chemicals, but not much better than other fish, despite their reputation. Sharks seem to need much more than even 100,000 amino acids per liter—maybe as much as several billion per liter—before they sense the proteins. So the drop of blood in the ocean is unlikely to attract a shark.

But what about the old traditional Olympic-size swimming pool? Here the shark's chances are a bit better. A drop of blood in 2.5 million liters yields about 320 billion amino acids per liter of water. That is likely well within the shark's capabilities, and even five Olympic-size swimming pools might not be too much. (The closest a shark has ever come to actually being in the Olympics was during a race against US Olympian Michael Phelps. The shark was a computer simulation; Phelps still lost.)

Of course in the real ocean, a drop—or even a gallon—of blood wouldn't spread out evenly and smoothly in all directions. Water is tossed this way and that because of wind and currents, so the experience for a passing shark would be quite unpredictable. It could swim right into a patch of blood, or it could swim by detecting nothing.

Did You Know . . . The hammerhead shark is one of the most peculiar beasts of the sea. Its head is extremely flat and wide, extending out beyond the body on both sides. Its eyes sit at the very ends of the head and its nostrils are far apart as well. There's been a lot of interesting work done on the hammerhead's vision, showing that having their eyes out on the ends of their extended head gives them 360-degree vision and, amazingly, also good stereo vision. But what about smell? It's possible their widely spaced nostrils make it easier for them to sample odors in the water. The old idea would have been that they can tell which nostril has encountered the more concentrated odor plume, but the new idea might be that they're able to detect time differences in odor arrival better than other sharks—all in order to find the source faster.

The fact that sharks are not hypersensitive to blood and other amino acids makes sense. Some scientists have suggested that being more acutely sensitive than they already are might overwhelm them. The ocean, after all, is filled with living things; amino acids are leaking out everywhere, not just from wounded prey.

So the lesson here? If you cut your finger while in the ocean, you don't have to worry so much. But when swimming in a shark-infested pool—that's another matter altogether!

How do mosquitoes bite without my noticing?

SOMETIMES YOU CAN FEEL a mosquito land on your arm or ankle and occasionally you feel a sharp pain, but most often you only realize later what's happened because of the itch. How does the mosquito get away with that?

It all has to do with its amazing landing and takeoff techniques—and its precision feeding.

But before a female mosquito (they're the ones that bite) can drink your blood, she has to find you. To do that, she homes in on a combination of your breath, odor, humidity, and color.

Once a mosquito has you in her sights, she has to land lightly enough that she doesn't draw your attention. She weighs only about

My buzz is worse than my bite.

2½ milligrams (9/100,000 of an ounce), but the mechanoreceptors in your skin, the nerves that sense pressure, are extremely sensitive. So she reduces her impact and evades their detection by hovering and distributing her weight over her six legs as she sets down.

That's only the first step. Now she has to feed without detection. The mosquito uses heat-detecting elements on her antennae to lead her to a place on your skin with a blood vessel just under the surface. She doesn't "bite," however, but instead uses her needlelike proboscis, a six-tool package in a sheath, to drill through the skin. Four of these tools are serrated "stylets" that both grip and pierce the skin. The other two are extremely thin tubes, which inject saliva into the blood (to minimize clotting) and check for the presence of chemicals that signal a blood vessel nearby. These tubes are amazingly flexible—videos of a mosquito's proboscis fishing around for a blood vessel under the skin of a mouse reveal it bending at right angles. Several seconds and multiple probes might be required before a mosquito finds what she's looking for. Then she either inserts her tubes right into the vessel or punctures it, causing a hemorrhage, which she then laps up.

Science _Fact!_ The mosquito stops feeding because a specific nerve detects that her abdomen is distended, and she withdraws her blood-sucking apparatus. If that nerve is severed, she won't stop taking blood until her abdomen explodes!

From the time of landing to getting ready to leave, a mosquito might be on your skin for as long as a couple of minutes or more. The fact that we are so often unaware of this is a testament to the delicacy of her feeding equipment.

But there's one more big challenge. If a mosquito has been undisturbed and is fully fed, she might weigh double what she did when she landed, making it much harder to take off unnoticed than it was to land. Now she must push down with her legs to become airborne, but if she pushes with too much force, you'll feel it, and *smack*! All that hard work is down the drain.

Nonetheless mosquitoes do escape, so they must be doing something right.

Dutch scientists used extremely fast film to capture the departure techniques of mosquitoes who had just imbibed a gut-full of blood, and as you might expect, the insects do everything they can not to leave an impression. To keep the downward force below the sensitivity threshold of the skin mechanoreceptors, the mosquito starts to beat her wings at a furious frequency 30 thousandths of a second before takeoff. The wings are taking over about 60 percent of the takeoff force, and of course they push on the air, not on the skin. She still has to generate 40 percent of the push with her legs, but by pushing a little more lightly than usual, but for longer, the mosquito can get airborne without triggering your nervous system.

I know he's around
here somewhere...

A gentler takeoff should mean a more labored takeoff, and it does: Most of the well-fed females were indeed a little slower to gain altitude and maximum velocity than unfed females. At the same time, fed females made several changes to the way they flew. Their wings beat at the same extreme pace, something like 700 per second, but the mosquitoes altered the angle of the wings and the length and orientation of the wing stroke, all in an attempt to boost acceleration enough to make their escape, even though packing a body's worth of blood.

The combination of less force on the skin with reasonable acceleration means that fully fed females can get away much of the time without our even noticing that we've just been dinner.

Science Fact! *You might realize that you've been bitten only when you begin to feel itchy. But that is really your own doing. The mosquito injects anticoagulants into you to ensure that the blood flows freely, but your immune system recognizes these substances as foreign and reacts against them, causing the swelling, redness, and itch.*

Is it true that spiders could eat every single human on the planet?

VERY RARELY DO YOU GET A SCIENTIFIC REPORT with an absolutely jaw-dropping title like this: "An estimated 400–800 million tons of prey are annually killed by the global spider community."

Not sure how I'm going to get out of this bind.

Can it really be true? And if so, how would you figure that out?

Well, first, you'd have to know how many spiders make up that "global community." And second, you'd have to work out how much those hungry arachnids eat.

There's no doubt that Earth has spiders aplenty. More than thirty thousand species have been identified, from those in the benign-sounding cobweb spider family to the nightmare-inducing wolf spiders.

And a good many of these spiders are likely keeping you company right now. One study of fifty houses in North Carolina found evidence of fifteen different families of spiders in those houses, including jumping spiders, orb weavers, and sheet-web weavers. Every one of those fifteen was represented in the living rooms, dining rooms, and kitchens in those houses, while the bedrooms lacked only the so-called sac spiders. Cobweb spiders were found in every house.

Also it's apparently true that as you read this, there's likely at least one spider as close as, say, four feet (a little more than a meter) to you. But since you probably haven't noticed it, you can appreciate how hard it would be to account for all the spiders in your vicinity.

There are estimates that some environments are home to more a thousand spiders per square meter, but how can scientists calculate the worldwide population? Here's how the authors of the latest scientific estimate, Martin Nyffeler and Klaus Birkhofer, did it.

Over the years there have been more than sixty estimates of the numbers of spiders in various environments around the world. The estimates are based on spiders found in traps or counted on webs or even collected by vacuum tubes. Nyffeler and Birkhofer combined these estimates, being careful to adjust for each environment (because spider populations in forests and grasslands are denser than those in the desert or on the tundra).

Next, they were able to estimate the annual prey killed by spiders by using both the spiders' food requirements per body mass and data from actual surveys of spider habitat. Even insects left in webs and not eaten were considered prey.

Their final estimates depended on some assumptions, like that a typical spider consumes about 80 percent of its prey (sucks out the juicy bits and throws away the hard exoskeleton), and overall, every day, it consumes an amount of prey equal to 10 percent of its body mass.

They came up with an estimate for the global spider population of 25 million metric tons. Those spiders consumed anywhere from 400 million to 800 million metric tons of prey per year. Take those numbers, change them to pounds, and you have 55 billion pounds of spiders eating 1.76 trillion pounds of prey. Those numbers compare roughly to the annual consumption of meat and fish by humans or whales.

Spider prey by contrast is mostly other spiders and a wide range of insects, including moths, butterflies, wasps, bees, and ants. (Even though there is a creature called the bird-eating spider, *Theraphosa blondi*, apparently it hardly eats birds at all.)

The scientists Nyffeler and Birkhofer are not responsible for the claim that spiders could eat every human in a year; that piece of good news appeared in several publications around the world. The argument was straightforward: the biomass of all adult humans on Earth today is estimated at 278 million tons. If you add about another 70 million for the kids, you're still not anywhere near the amount the world's spiders consume every year. Providing they'd be willing to switch diets, they could start on New Year's Day and probably finish off the entire human species before summer. I don't know about you, but I'm glad that, so far, spiders appear to be just sampling the human tasting menu (they do bite us), but that's all.

Why don't woodpeckers get headaches?

MAYBE THEY DO, AND THEY'RE JUST INCREDIBLY STOIC.

That's unlikely, though, isn't it? Animals do feel pain, after all, and it's hard to imagine why a woodpecker would bang its beak into a tree trunk with extreme force twenty times every second if it hurt.

I use the word *force* because the woodpecker's beak can be traveling up to 5 or 6 meters per second when it is stopped dead by the wood. That deceleration force is 1000 g's—more than a thousand times the force of gravity. And the bird may do that 12,000 times a day! So how does that not cause a headache?

Sometimes I feel like I'm banging my head against a wall.

Well, the woodpecker's head is much smaller than ours—and that should make a difference. A smaller head will have more surface relative to its volume, so relatively more impact area and more cushioning. It's a factor, but not the whole story, and here's why. A typical woodpecker brain might weight 2.5 grams; the human brain, 1,400. Do the math on that, and the numbers suggest that the woodpecker brain should be able to resist, on the basis of size alone, forces sixteen times greater than those a human can withstand.

Humans will lose consciousness somewhere between 4 and 6 g's acceleration/deceleration. Woodpeckers, based on their much smaller brain, should theoretically be able to withstand 65.5 g's before they experience G-LOC (g-force-induced loss of consciousness). Actually, though, they can withstand much more—like 1000 g's. Whatever is protecting the woodpecker is more than just having a small head. Size alone would not prevent them from being knocked out.

One thing that helps is that the impact lasts a very short time. The longer an impact lasts, the greater the potential for damage. In a car crash, if the passenger is not wearing a seat belt, colliding with some part of the car, like the dashboard, will last anywhere from 3 to 7 thousandths of a second. In the National Football League, players suffer concussions from impacts lasting 15 thousandths of a second. Woodpeckers, by contrast, experience brief impacts—less than 1 thousandth of a second.

Then this bird has a set of cool anatomical features that allow it to hammer away, even at metallic surfaces, for extended periods of time. The beak experiences the impact first, but the forces of impact flow through it to the skull, dampening those forces to the point where the brain is not injured. Some of the force is channeled downward to the base of the skull. The remaining decelerating force immediately spreads through the beak to the front of the skull, particularly the circle of bone, the orbit, surrounding the eye. This might explain why woodpeckers close their eyes just before impact. (One experiment showed that there were significant differences of force transfer depending on whether the lower tip of the beak hit before the upper tip or vice versa.)

If the bird's brain was floating in its skull, as ours is, then the kinds of impacts created by hammering on a tree would be expected to jerk the brain around inside the skull, potentially damaging it. The woodpecker is protected from that in two ways: The bone at the front of the skull is spongy and more impact-absorbent than the rest of the skull, but there is also relatively little fluid surrounding the brain inside the skull, which makes bouncing around much less likely.

The woodpecker's brain is packed in there tightly, and is even positioned so that its greatest surface area is facing the front where the forces come from. Distributing those forces over a larger area helps.

Most of these adaptations are simply embellishments or evolutionary tinkering with the structures that all birds have in one form or another. But the woodpecker does have one astounding feature that plays a leading role in all this: its tongue. If you can envision this, the woodpecker's tongue starts in the mouth as you'd expect, but then toward the back of the beak, it splits in two. These two strands dive down under the brain, then up the back, curling over the top of the head, then down toward the beak again. This hyoid apparatus (more than a tongue, really) is cartilaginous, much tougher and stronger than the typical tongue, with a hard core and a more pliant exterior. While it's not exactly clear how much of the strike force is absorbed by this structure, some have suggested it acts like a "safety strap" or even a seat belt, stabilizing the brain by making the entire head more rigid than it would be otherwise.

Some research groups have justified their work on woodpeckers by pointing out that anything we learn about these birds might be applicable to safety helmets for humans. But it would be challenging to fashion this unique set of adaptations into a helmet for a human. We're not likely to ever get our tongues to wrap around our skulls the way woodpeckers can.

The Amazing Skull of a Woodpecker

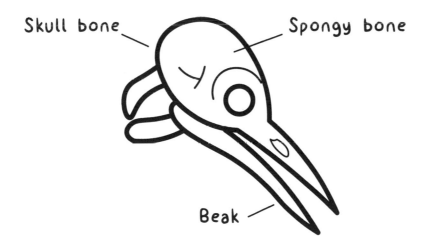

Skull bone

Spongy bone

Beak

Is it true that only half of a dolphin's brain sleeps at any given time?

You don't usually think of sleep as dangerous, but it can be, especially for wild animals. If they don't get enough, they can die. But at the same time they face a number of risks while sleeping. It looks as if dolphins (and as it turns out, many other animals and birds) have solved the problem by catching up on sleep one brain hemisphere at a time.

John C. Lilly, one of the original dolphin experts, observed in 1964 that sleeping dolphins closed only one eye at a time, first one for a while, then the other. He suspected that the closed eye was directly connected to a sleeping hemisphere (each side of the brain generally controls the opposite half of the body). Lilly's intuition was later proven right by electrical recordings of activity in both sides of the dolphin's brain. The bottlenose dolphin, for instance, logs eight hours of sleep daily, four hours on each side of the brain. Also dolphins and their relatives— porpoises, killer whales, belugas, and likely many others—seem not to experience the lighter, rapid eye movement (REM) sleep during which dreaming takes place.

Why do dolphins and their kin do this? Well, to breathe, they have to come to the surface of the water regularly. Indeed, they and their fellow cetaceans swim constantly, probably driven by this need to surface. If only one hemisphere is necessary to control swimming movements, then the other hemisphere can sleep—a solution to the problem of breathing while sleeping, two life necessities. Dolphins in pods have also been seen orienting themselves to keep the one open eye on the others, suggesting that single-hemisphere sleep also enables the pod to stay together. (You'd think that the eye that's open should be looking away from the pod for approaching predators, but that is apparently not the case.)

But it's not just dolphins. Seals, manatees, and a long list of birds, including chickens and pigeons, all spend some time sleeping with only one hemisphere of their brain, while the other remains awake. And they're not doing it so they can breathe! (This happens only with deep sleep, however: when any of these animals enter REM sleep—and unlike dolphins, these animals are capable of REM sleep—both hemispheres are involved.)

Science Fact! *Among birds, chickens take the hemispheric specialization to a new level. When sorting through gravel on the ground to pick out edible seeds, chickens use their right eye/left brain combo. By contrast, scouting the environment for danger engages the left eye/right hemisphere. This is a convenient pairing, because the bird can cock its head to focus the right eye on the ground to search for food while its left eye scans the sky above.*

Frigate birds sleep half a brain at a time so they can keep fishing. They're not adapted to touch down on the water, so they are airborne for days on end, patrolling the ocean looking for fish at the surface. But during that time, they need to get a little shut-eye. It sounds unbelievable, but these birds can fly with one hemisphere, or even both, asleep, although the most sleep time they seem to manage is about forty-five minutes every twenty-four hours while flying. (They catch up by sleeping ten hours per day when on land.)

Did You Know . . . Most sleep research has been done with humans (of course, at least to us, we are the most interesting species), and while we have discovered quite a bit, why we sleep as much as we do is still somewhat of a mystery. The known benefits include the consolidation of memory, especially of events from the day before, and accelerated brain development in infancy. But that short list seems far from complete.

How important is our sleep? In 1965, in order to achieve the world record, seventeen-year-old Randy Gardner stayed awake for about 11 days (264 hours). By day five, he was already hallucinating, his attention span had been dramatically reduced, and he wouldn't talk unless spoken to. Toward the end of the 11 days, his mental functions had declined even further, even though over the last few days of his ordeal he was likely experiencing momentary brief microsleeps—it just wasn't possible for him to stay completely awake.

Gardner recovered without incident after a sleep of 14 hours, followed by another of 10 hours, but the effects on him during the marathon had been so pronounced that *The Guinness Book of World Records* removed voluntary sleep deprivation from its list to discourage further attempts. People *have* died from the effects of sleep deprivation, like the British pilots who fell asleep and crashed returning to England from bombing runs in World War II. But even in a safe environment, the effects of deprivation are thought to be extremely dangerous. Rats, for example, will die if prevented from sleeping: apparently their immune systems collapse.

I'm so tired I could die.

A number of animals divide their sleep between hemispheres to protect themselves from predators. Like dolphins, northern fur seals, who live in the Pacific Ocean, sleep in the water, but have to keep partially awake in order to breathe. They do this by lying on their side, with their nostrils out of the water, one half of their brains moving the flippers on their submerged side while the other nods off. When they flip over, the brain hemispheres switch roles. Interestingly, however, fur seals also sleep one hemisphere at a time on land, where breathing is not a challenge, suggesting that they do so to keep a lookout for predators.

So dolphins aren't the only creatures that sleep with only half a brain. But do humans ever have one hemisphere sleeping with the other awake? No, not really. But when sleeping for the first time in an unfamiliar place, one hemisphere does sleep more lightly than the other and is more easily roused. It's not awake, but it is keeping watch for us in its own way!

ZZZZZzzzzzzzzzz...

Why are a grasshopper's ears located on its abdomen?

WE HUMANS THINK IT'S WEIRD that grasshoppers have ears on the sides of their bodies, and that their relatives the crickets have ears on their legs. But in fact we and all other vertebrates are the weird ones, because we are limited to having ears on the sides of our heads. Insects, in contrast, can have ears almost anywhere: some moths have an ear in their mouthparts; horse-flies, where their front leg joins the body; praying mantises, between their back legs. Pick a place on an insect body, and you might just find a species with an ear located there. Or two, or in at least one case, twelve!

In appearance, insect ears are very different from ours, mostly because they don't have much of an appearance. No complex twists and folds like our ears: generally just places on the insect's body where the hard-shelled exoskeleton is thinned, sometimes to the point of transparency, but always to the point where it can vibrate.

Those places on the body, wherever they are, are analogous to the human eardrum: structures that move back and forth in tune with the incoming waves of sound.

Did You Know . . . Since insects have ears everywhere, the question might be flipped: Why do humans, as well as all other vertebrates, have ears on their heads rather than, say, their knees? All animals with a backbone have ears located on the head, behind the eyes and above the jaws. That remarkable consistency can be attributed to the evolution of the hearing apparatus. It, like most of our body systems, had to come from somewhere, and in vertebrates, it has apparently been cobbled together (over hundreds of millions of years) from embryonic structures associated with the head.

There are a few reasons why insects have ears on various spots of their bodies. Insect ears evolved from structures that determined the position in space of everything from the insect's body to its legs to its antennae. In evolutionary terms it's relatively easy to convert these into ears, and because such sensory organs can be found almost everywhere on the insect, their gradual specialization into organs of hearing make it possible for insect ears to be many places other than the head.

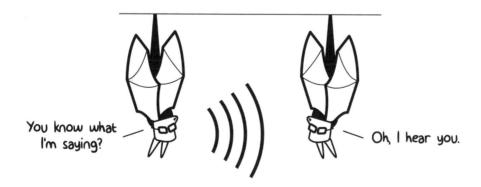

Another reason why insect ears have ended up on the body parts they do is because many species of insects need to be able to determine where sound is coming from. Whether it's a female cricket trying to find the source of a mating call ("call" in this case can be the male cricket rubbing his legs together!), or a moth needing to know where that threatening bat sonar signal is from, a quick and precise determination of direction is essential.

One way of doing that is to compare either the time of arrival or the intensity of the sound coming from the right or left. The farther apart a pair of right and left ears are, the better that discrimination. Insects that have ears widely separated, the way they would be if they were located, say, on their legs, would have an advantage over those with ears on either side of their heads (although this is theoretical, because there are apparently *no* insects that have ears located like ours). The direction from which high frequency sounds originate is relatively easy to discriminate by time or arrival. This doesn't work as well for low frequencies, so insects that need to detect *those* have ears in places where both the inside and the outside of the eardrum are impacted by the sound.

Science Fact! *Many species' hearing is tuned to precisely the frequencies of sounds that other members of their species make. Male crickets "chirp" to attract a female. The female cricket's ear is tuned to that sound and not many others. Moths' hearing is tuned to the ultrasonic hunting sonar of bats, and when they hear it, they will take evasive action by folding their wings and dropping like a stone.*

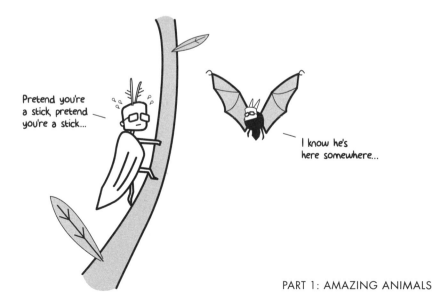

And finally, there probably isn't even room in an insect's head for a pair of ears, or even a single one. Those insects that have sophisticated hearing systems use a combination of sensory cells, air-filled tubes, and associated structures. Their body cavities are likely the only feasible place to locate all of these parts.

So put your hands together for the insects: they're tiny, have little space to work with, but are equipped with an amazing collection of ears crafted for them by evolution.

What was a dodo?

IF YOU HAVE EVEN A VAGUE IDEA OF WHAT A DODO WAS, you probably think it was some kind of now-extinct bird that lived in an exotic location and was, well, really stupid.

All of that is true—except the "stupid" part. But we'll get to that later.

The interesting thing is that even the experts still have a whole lot of unanswered questions.

The dodo-human relationship was pretty brief as such things go. The Dutch first chose the island of Mauritius (that's the exotic location in the Indian Ocean) as a stopping-off point for their ships in 1598 and were introduced to the dodoes. The birds were an instant curiosity (or dinner) to visiting sailors. "We also found large birds, with wings as large as a pigeon's, so that they could not fly and were named penguins by the Portuguese. These

I'm no dumb-dumb.

particular birds have a stomach so large that it could provide two men with a tasty meal and was actually the most delicious part of the bird." (They ate the stomach?)

The last living dodo was reported in 1662. In just six decades the birds were wiped out, although of course a few individuals might have hung on for a decade or two in remote parts of the main island or on one of the tiny companion islands.

There is much that remains a mystery about the dodo, including what it looked like. But it's certainly gone: we have more skeletal remains of individual dinosaur species than we have of dodoes. Pull all the dodo remains in the world together, and here's what you have: a lot of loose bones, impossible to assemble into skeletons of single individuals. There is, in fact, but one complete skeleton. And preserved dodoes? One foot with almost no flesh left on it and one dried head with almost no feathers. That's it.

The written descriptions and paintings of the bird are our only compensation for the lack of actual remains. But experts today put little confidence in most of the writings (even though it's been claimed that there was more written about this bird than about any other), because we have no way of confirming how accurate they were. For instance, many of the anatomical details described, like the colors of the birds' bills, can't really be substantiated—there is only the one bill in existence. Other than the single foot, no other body parts have survived, so claims about, say, the color of the tail feathers simply can't be counted on. Many of these written descriptions were likely constructed from rumors, gossip, or paintings.

Even the paintings are undependable because most appear to depict captive birds. Many dodoes were seized to take back to Europe, but the best estimates are that only three or four actually made it there alive. Many of the artworks featuring dodoes were painted not in Mauritius, but in France. Some might have been of live birds, but many were of stuffed ones. The quality of the taxidermy would determine how representative of the living bird the painting was.

Most of the paintings depict a fat bird. But were dodoes really that fat? (Or as one writer put it, was the dodo a "stupendously overweight idiot of a bird"?) The most famous portrait was made by Dutch painter Roelandt Savery in around 1626 and was later copied by Sir John Tenniel in his illustrations for Lewis Carroll's *Alice's Adventures in Wonderland*. It makes the dodo look like it had been stuffed *before* it died.

Dodoes were definitely big birds, apparently about three feet tall, and might have weighed anywhere between 25 and 30 pounds, but even though some estimates put them at as much as 50 pounds, Savery's bird might have had to diet even to make that weight.

A team of French scientists recently took a second look at the "fat" dodo by measuring and analyzing sets of leg bones to use as an estimate of body weight. There's a reliable relationship between the two, and the dodo leg bones yielded an estimate of average body weight of 10.2 kilograms, or just over 22 pounds—not 50! Those stories of entire ship's crews gorging themselves on a few dodoes might just have been exaggerations.

The leg bones tell us more about the birds than just its weight. The dodo was flightless, having wings so small that at best they could have been used in mating displays, meaning that they had to depend on those legs to escape predators. Even though they are often described as clumsy and slow, examination of the leg bones reveals that the legs were large and the kneecaps thick, suggesting that the birds could move with agility and even some speed—maybe using those vestigial wings for balance. In fact, although dodoes have been fat-shamed for centuries, a number of Dutch sailors reported that these birds were very fast afoot.

And now for what we know about their extinction. The idea that somehow dodoes were a type of evolutionary dead end and thus deserved their fate doesn't seem to make sense. Detailed studies of their bones reveal that they were active, agile, and beautifully adapted to the Mauritian environment.

The more popular theory, however—the story that has been peddled for almost three hundred years—is that they died out because of their stupidity. Supposedly they were just too dim-witted to avoid being slaughtered by those Europeans. Yes, there are countless tales of the birds placidly allowing themselves to be hunted by the humans arriving on the shores of Mauritius. But that story needs to be revised.

Lack of familiarity with humans doesn't make an animal or bird stupid. It usually means that such creatures will not automatically flee for their lives when humans first appear. This is often the case when humans first arrive on an island, like Mauritius.

Moreover, it's very unlikely dodoes were hunted to extinction by humans anyway. For one thing, throughout the 1600s there were never more than a couple hundred people on Mauritius, often many fewer, and all of them lived on the coast. The interior of the island was wild and unsettled. Also a refuse pit on Mauritius full of bones from slaughtered animals, 10,000 of them, included not one dodo bone. And there were several negative reviews of the tastiness of dodo meat.

Humans did, however, contribute to the dodo's demise by bringing in other animals: goats, cattle, and especially pigs and rats. Flightlessness was irrelevant to the survival of the dodo before these animals arrived, but it was a definite disadvantage in the face of the influx of invasive species that killed and ate the birds and their eggs.

So what we *do* know about dodoes is that whether they were fat and stupid or trim and quick-witted, they would still be roaming Mauritius if we hadn't happened along.

How do monarch butterflies migrate from Mexico to Canada and back?

THIS IS A TRICKY QUESTION THAT ACTUALLY DOESN'T HAVE AN ANSWER, because the monarchs that return to Mexico have never *been* to Mexico. They were born in North America in the summer. Their great-grandparents were the ones that left Mexico the previous spring. So memory, at least in the way we think of it, isn't the answer. But without a doubt the monarch's travels are the most amazing migrational journey undertaken by any insect.

It begins, for most monarchs, in forests of oyamel firs at high altitudes in central Mexico. (The decline of those forests is one reason monarchs are endangered.) There, monarchs cluster in the millions from fall to spring, when they rouse themselves, mate, and then take off north to Texas. Upon their arrival in Texas, the females lay their fertilized eggs (they can produce their

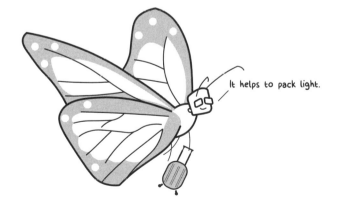

It helps to pack light.

weight in eggs!) and die. The eggs hatch and the striking yellow and black caterpillars begin to feed on milkweed plants, their sole source of nourishment. These larvae eat continuously and go through several stages, gaining as much as two thousand times their original weight. They then finally enter the pupal stage.

Science _Fact!_ *The pupal stage itself is amazing. In the pupal case, or cocoon, the internal organs of the monarch are broken down into a kind of organic soup and then reorganized, changing a caterpillar into the orange and black butterfly that we're familiar with. The basic structure of some organs is preserved through this dramatic change.*

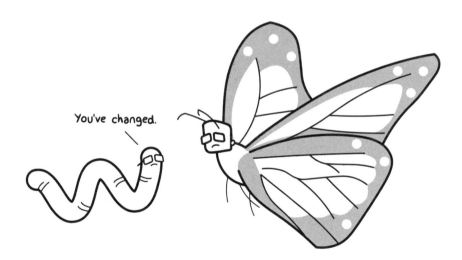

You've changed.

The new adults emerge from the cocoon and dry their wings. Then they're off—north. This first generation of adults is not long-lived: they'll feed on nectar, fly, mate, and lay eggs over a period of three weeks or so. Then they die and the process is repeated. By this time, the leading edge of the monarch migration might have reached midwestern states like Missouri or Nebraska. The whole feed'n'fly'n'reproduce happens again, and now there are young butterflies flying into southern Ontario and Minnesota. That is roughly the northernmost point on this journey, although monarchs have been spotted farther, in Edmonton in the west and even near James Bay.

Once this generation of butterflies has reached the northern United States or Canada and laid their eggs, things change dramatically. These newest adults will live not a few weeks, but a few months. They will shut down their reproductive apparatus for that time. They will become incredibly gregarious, roosting in trees in the thousands. And most important, they will now undertake an incredible 4,000-kilometer (2,500-mile) migration south across the United States to the forests from which their great-grandparents came—never having been there before.

Did You Know . . . Monarch butterflies are poisonous! The caterpillars feed exclusively on milkweed, which contains toxic chemicals that the caterpillars can tolerate but predators, especially birds, can't. The chemical makes them sick to their stomach. This is likely why the caterpillars are anything but camouflaged: their yellow, black, and white stripes stand out against the green milkweed plants, in effect advertising their presence. Any animal that has had an unpleasant experience with them will leave them alone.

How would you go from Canada to Mexico, never having been to Mexico previously, if you didn't have Google maps? You would need a map of *some* kind and a compass. And all along the way you'd have to confirm how far you'd gone and where you were. The monarch does all of that in its head.

First, how does the insect "know" which direction is south? ("Know" is in quotation marks here because they're probably not thinking about it, just doing it.) The sun is the best indicator of south, rising as it does in the east and setting in the west. But there's a problem with that. To head south in the morning, a butterfly has to keep the sun on its left—in the east. But by the end of the day, if it kept doing that, it would be heading back north instead of south. That won't work. Therefore it's not enough to use the sun by itself; the sun's position has to be coupled with some sort of clock. That combination can be used to fly south by comparing, minute by minute, the position of the sun with the time.

We're talking here about an insect with a brain the size of the period at the end of this sentence. That means there's not a lot of gray matter available to be applied to this problem. Nonetheless monarchs can do that—and more. When it's too cloudy to read the position of the sun accurately enough, the monarch turns to two other guides: polarized light from the sun and Earth's magnetic field.

Brain of a butterfly or end of a sentence?

As sunlight enters Earth's atmosphere, it gets bumped around—*scattered* is the term—forming circular patterns in the sky, centered on the sun, that the butterfly can see, even though we can't. One tiny corner of the monarch's eye is sensitive to the polarized light around them, and this gives monarchs additional clues to the location of the sun, helpful on a cloudy day (although a patch of blue sky is necessary for this to work).

But the butterflies' antennae are also sensitive to Earth's magnetic field, which not only provides information about direction, but also how far north and south they are.

Science Fact! *The only time human beings are sensitive to polarized light is when we see the glare coming off the surface of a lake or a road. We put on polarizing sunglasses to eliminate those horizontal waves.*

All of these intricate calculations are taking place inside the tiny space of the butterfly's head, and with some parts of this system, only a few neurons are employed to guide the insect. And remember in the spring the map/compass directs the butterflies to fly north, but in the fall, it guides them south. This 180-degree change of direction has to be established in the brains of the generation born in the northern United States or Canada.

The genome of the monarch butterfly may be our best shot at really understanding how all of this works. For one thing, even though they are visual creatures, monarchs have the same genetic repertoire for odors as insects that depend on messaging by odors called pheromones do. Pheromone communication might explain how, on their fall migration, monarchs maintain coherent flocks numbering thousands. Closer inspection of the genes for hormones will likely clarify how it is that the last generation lives so much longer than the previous ones. And because there are some populations of monarchs that don't migrate, such as those on the Hawaiian Islands, genetic analysis should be able to pinpoint which genes are responsible for the migratory urge.

But until that genetic work is done, the monarch migration is still a mysterious wonder.

Migration is for dummies.

History Mystery

Has Earth ever been visited by aliens?

"Has Earth in the past been visited by intelligent beings?" Robert K. G. Temple's 1976 book, *The Sirius Mystery*, asks just this. He used the next 400 pages to answer that in the affirmative: yes, it has.

Temple noted that anthropologist Marcel Griaule's interviews with the Dogon people of the African country of Mali from the 1930s to the 1950s suggested that they had scientific knowledge of the star Sirius far beyond their capabilities, which therefore meant perhaps another species had helped them discover it.

That Sirius held meaning for the Dogon isn't mysterious: First, it is the brightest star in the sky (sometimes visible in daylight), and thus impossible *not* to notice. Second, it has been an important star for many cultures, especially that of ancient Egypt. Egyptians paid close attention to its first rising, just before sunrise on July 19 (at the latitude of Cairo), because it heralded the flooding of the Nile. But the Dogon's detailed knowledge of Sirius, as reported by Griaule and further embellished by Temple, was extraordinary.

Apparently the Dogon were aware that Sirius (*sigi tolo*) had a dim companion star (now known as Sirius B); that this companion, which they called *po tolo*, was made of super-dense material (*sagala*); and that it orbited the main star, Sirius A, every fifty years. They were also aware of a third star in that system, and they had apparently known all this for several centuries. There were other curious facts, too—notably that they were aware of the rings of Saturn and the four moons of Jupiter.

These revelations were a multilayered puzzle. Sirius B cannot be seen without a telescope, and a powerful telescope at that. It can't be seen with the naked eye, not only because it is very small, but also because Sirius A is so blindingly brilliant it washes out the feeble light from Sirius B. It was only discovered in 1862 by astronomer Alvan Graham Clark using a huge 18.5-inch (47-centimeter) aperture refractor telescope, the biggest telescope in the United States at the time.

How the Dogon knew of Sirius B's existence centuries ago without telescopes is one part of the puzzle. But their description of the companion as being super-dense is uncannily accurate as well: Sirius B was confirmed as a white dwarf star in 1915. White dwarves are the remnants of a stellar collapse, and they pack the mass of the sun into an object the size of the earth. And just to complicate things further, the Dogon's knowledge about the smaller star's orbit is accurate, too: the orbit period is actually 50.1 years. So far there is no evidence of a third companion star that the Dogon referenced.

Griaule the anthropologist was dumbfounded by these pieces of information. Indeed, while Dogon astronomy was said to be centuries old, they simply didn't have the technology to know about a distant, super-dense invisible star traveling in a fifty-year orbit around a giant bright companion. Robert Temple, however, claimed he knew exactly how they became aware of all those. In *The Sirius Mystery*, he argued that aliens from the vicinity of Sirius came to Earth and passed this information on to the ancient Sumerians, who then conveyed it to the ancient Egyptians, whose knowledge somehow filtered across the African continent to the Dogon.

If you're unwilling to accept that story, you have to come up with an alternative explanation. For instance, could some exceptionally sharp-eyed members of the tribe have actually seen the tiny companion Sirius B? Unlikely to impossible, not just because of the overwhelming brightness and proximity of Sirius A, but also because the intrinsic dimness of B is well beyond the abilities of even the sharpest-eyed humans gazing from high altitudes where the air is clearest.

Even if Sirius B had been spotted, there would be no way to know that its orbit was fifty years (unless this sharp-eyed individual was extremely patient) nor that it was extremely dense.

So we're left with two explanations: Temple's suggestion that aliens from Sirius brought all this knowledge (indirectly) to the Dogon, or the explanation favored by astronomers, which is that the information traveled to the Dogon by way of France.

There was a lot of chatter about Sirius's companion star and its incredible density in France in the 1920s, including reports in newspapers like *Le Monde*. Over the next couple of decades, Europeans, including many French, visited the Dogon, and some Dogon tribesmen fought in the French army in World War I. There had been French schools in the area since the early 1900s and missionaries aplenty. Sirius might have been part of *their* conversations.

That does raise a question, though: even if the Dogon were told tales of Sirius (and they might have been interested because it was the only star they had named), would they actually have inserted that information into their tribal beliefs?

There are some examples of modern knowledge being grafted onto traditional wisdom. In Papua New Guinea in the 1950s, photomicrographs of spiral bacteria were shown to local people and became part of their beliefs about how malevolent spirits (as represented by wavy lines drawn in the sand) could enter the body and cause disease. Now, in one case we're talking about Papua New Guinea, and in the other, Mali. These cases might not be comparable, but this scenario makes more sense than aliens.

The rumor that we are malevolent spirits contains only a germ of truth.

But there is also the chance that the Dogon did not, in fact, possess scientific knowledge of Sirius at all. Anthropologists have questioned several of Griaule's accounts of what he had been told. Walter van Beek argued that when he lived with the Dogon, few knew anything about Sirius's being a double star, nor anything about its mass or its orbital period. The only Dogon who had scraps of this knowledge were part of a small group whom Griaule had spent his time with. Van Beek noted that many of the Dogon told him that they had learned about the star from Griaule himself. Van Beek has his critics, too, but at this point it's fair to say that we're not really sure how prominent Sirius was in the Dogon belief system before the Dogon were visited by Europeans who knew all about Sirius.

Temple's book, which hinges on the Dogon's ideas about Sirius being several centuries old, is weakened by doubts about those claims. But really his book is weak enough on its own. For instance, if aliens really did arrive from Sirius and tell of Jupiter's four moons and Saturn's rings, why didn't they get those facts right? Jupiter has, at latest count, 79 moons, not four (but four were all that were known in the early twentieth century). Saturn is not the only planet with rings—there are three others. Nor is Saturn the outermost planet in the solar system. Why didn't the aliens reveal this, seeing as how they must have passed right by them on their way here?

What could move this story ahead? The discovery of the third star of the Sirius system? There have been reports of its existence—subsequently disproven—but even if one was discovered, there had been talk of a third star early in the twentieth century, so it wouldn't prove anything.

Your choice: aliens or not!

Part 2
The Body

Why do people stick out their tongue?

You probably know what it means when someone gives you the finger or sticks out their tongue at you. But did you know that when the tip of someone's tongue just peeks out from between their lips, they are telling you something, too? It's another kind of unspoken communication.

This sort of tongue-showing is easiest to envision if you think of young children concentrating as hard as they can on what they're doing. It could be something a little risky or mentally challenging—no matter what, it's absorbing. And more often than you might think, when a child's brain is locked in like that, the tongue appears, sometimes curled up over the top lip, sometimes firmly clamped between the teeth.

Researchers have confirmed this phenomenon in both children and adults. In one study at a nursery school, as children climbed a narrowing staircase of wooden blocks, then ventured out on a "bridge," observers noted their tongues poked out of their mouths. An experiment at the University of Pennsylvania showed that inexperienced pool players (who naturally have to concentrate hard on each shot) show their tongue much more than do experienced ones. But why?

Did You Know . . . Tongue-showing may have a long evolutionary history. Captive gorillas were seen to tongue-show when they were engaged in some sort of risky physical activity, like balancing on an upended wagon or swinging on chains, or in social interactions, like turning down an invitation to wrestle.

In one experiment conducted in Philadelphia, a young man sat at the top of a broad set of steps leading to an office building. As people climbing the stairs approached him, he stared directly at them. Some looked away, a few smiled, but seventeen out of fifty—almost exactly a third—tongue-showed (as recorded by a hidden observer). In the control experiment, where the man sat in exactly the same place on the stairs but read a book instead of stared, only two out of fifty tongue-showed as they neared him.

The experiment seemed to support the theory that when people tongue-show, they're saying, "Leave me alone," because they're either busy, preoccupied, concentrating, or just feeling antisocial. Strangely, neither sender nor receiver is conscious of the tongue's sending this message. But if neither person is aware of what's going on, how do we even know the signal works? There have been a couple of provocative experiments that have tried to determine if tongue showing really deters people from interrupting others.

One classic tongue-showing experiment was staged in a classroom. One student at a time (fifty in all) sat at a desk turned toward the back of the classroom; each was asked to take a reading comprehension test. The student was told that it was extremely important to answer the test questions in order. The supervising teacher sat at the front of the room, facing the student's back, and informed the student that he would be transcribing some audio and would wear headphones so as not to disturb the student.

Soon after starting the test, each student discovered that page three was missing. Having been warned of the importance of doing the test in order, students then had to interrupt the supervisor to let him know about the missing page. If they just swiveled their chair and called out, the supervisor, ears completely covered by clunky headphones, ignored them. They either had to yell or actually walk up to him, sometimes even touching his shoulder to get his attention. Again, a hidden observer timed how long it took the student to approach when the supervisor was either showing his tongue or not. If he wasn't showing his tongue, students took an average of 7.72 seconds to interrupt him, but when his tongue was showing, the average was 19.93 seconds. That 12-second difference, given the pressure of the test, was enormous. One student, obviously super-sensitive to tongue-showing, waited an extraordinary two minutes.

After the test section was over, each student was asked to sketch the supervisor's face. None of those who had actually been slowed by his tongue-showing added the tip of the tongue to their portrait, but many of them nonetheless described being very reluctant to interrupt him. Those students who hadn't seen the supervisor's tongue hadn't felt any such hesitation.

So no one involved seems to be consciously aware of it, but it seems to work. But this is peculiar, nonetheless. Why use the tip of the tongue to signal "Don't bother me"?

I speak in tongues.

Julia Chase, one of the main investigators of tongue-showing, has argued that the gesture hearkens back to infancy. She described a six-week-old baby who, when tired of nursing but too weak to pull her head back, would push the nipple out of her mouth. Charles Darwin had remarked on something similar, but it's pretty much anecdotal. The gesture might start within the first few weeks of life, and it may have a long evolutionary history, too. Gorillas have been seen to tongue-show whenever they're concentrating hard on whatever they're doing.

The puzzling thing about tongue-showing is the contrast between how common it is and how little research has been done on it since the 1970s. It makes you wonder how many other "secret" messages our bodies send that we have yet to discover.

Why are some people left-handed?

LEFT-HANDEDNESS IS CLOAKED IN MYSTERY, but the one thing that's certain is that only about 10 percent of the population is left-handed, no matter where you are in the world. In fact, a human population with more left-handers than right- has never been discovered; there's not even a population where left-handers make up as much as 25 percent. But as soon as you start wondering why that is, or what advantage or disadvantage left-handers have, you're on much shakier ground.

The consistency of that 10 percent number means it can't just be chance. It would be reasonable to guess it's genetic instead, but that's been very hard to prove. For instance, if handedness is ruled by genes, why don't identical twins (who have identical genomes) always have the same handedness? But not only do one in five pairs differ, but two left-handed parents can have right-handed children.

What's more, while a combination of two genes has been identified that seems to predict something close to the 10/90 split, one of the two genes is peculiar in that its effects are random: sometimes it can lead to left-handedness, sometimes not. If handedness is caused by genes and inheritance, the process is not straightforward.

And then there's the question of why this variation might exist. If nine out of ten people are right-handed, does that suggest it's better to be right-handed?

Well, there may be advantages today, but that's almost certainly because right-handers have designed the world. If you're left-handed and have trouble with right-handed scissors, that doesn't mean being left-handedness is a flaw—just that the scissors weren't made for you. No, if there's a disadvantage to left-handedness, it would have to show up in an environment that hadn't already been designed to suit right-handers, and it's pretty hard to find one. We would need to look back thousands of years.

Science _Fiction!_ It wasn't that long ago that left-handers were considered marked by the devil. Even as recently as the mid-twentieth century, schools in the United States, Canada, and the UK persuaded—forced, really—left-handed children to use their right hand. One of the excuses for this was that if you wrote with a fountain pen left-handed, you would smear the ink as your hand moved from left to right across the page. But there were even medical experts who went much further, labeling left-handers as awkward, stubborn, difficult, and willful. Such labels we now know to be ridiculous, but they were the last vestiges of much more serious disdain for left-handers in the centuries before. There are many biblical references to the right side being good and the left bad. The word dextrous, _meaning agile and skilled, comes from the Latin word for right. But the word_ sinister _is derived from the word for left._

You might wonder how we could even know whether people thousands of years ago were right- or left-handed, but there are some clever ways of figuring that out. For one thing, the arm that you favor is usually a little bit longer than the other one, and often, especially if you do hard physical work with your preferred hand, the bones will be a bit thicker. Using this approach, paleoanthropologists have concluded that our distant ancestors, like *Homo erectus* from a million years ago, were likely more right-handed than left-handed.

There are two other kinds of evidence suggesting that we have been predominantly right-handed for a very long time. For millennia, human ancestors made stone tools by holding a large stone in one hand and knocking sharp flakes off it with a stone held in the other. The patterns of flaking reveal which hand held the striking stone, suggesting it was the preferred hand. On that evidence, our close cousins, the Neanderthals, who dominated life in Europe for hundreds of thousands of years, seem to have been right-handers. More recently, the humans who created the beautiful cave paintings in France and Spain 30,000 years ago left evidence that they were right-handed, too. Many of their cave paintings were accompanied by hand prints, created by blowing pigment around a hand that was pressed against the cave wall. The majority of these stencils are left-handed, and most researchers think this was because the pigment, or some sort of tool, was being held in the right.

Science _Fact!_ *Handedness doesn't appear to go back as far as the common ancestor we share with chimpanzees. While modern chimps favor one hand for certain tasks, like using twigs to fish for termites, they prefer the other for different activities. If chimps don't have a preferred hand, it's unlikely our common ancestor did.*

Why did the right hand become the most commonly dominant one? Each hand is controlled by the opposite side of the brain. The left hemisphere, besides controlling the right hand, usually controls speech as well. Something like 95 percent of all right-handers have their main speech centers on the left side of the brain, and even 75 percent of left-handers do as well. Researchers have wrestled with the idea of somehow connecting handedness, speech, and toolmaking into one grand scheme, even suggesting that hand gestures are a bridge between physical movement and language. But so far there is no clear connection.

Did You Know . . . In the 1980s one scientific report suggested that, on average, the lives of left-handers were shorter than right-handers by nine years! While that sounded like terrible news for lefties, since then researchers have pointed out the bias in the study. The oldest subjects grew up in a time when children were often forced to change their handedness. The fact that the study found no eighty-year-old left-handers did not mean that all the lefties had died young—just that anyone born left-handed had had to become right-handed in their youth.

So what do left-handers have going for them? One of the most popular ideas is that left-handedness is an advantage in hand-to-hand fighting. There are a number of one-on-one sports with unusually high numbers of left-handers—like mixed martial arts, boxing, and fencing.

That might mean that in hand-to-hand situations, right-handers are not used to facing left-handers (there are so few of them), and they therefore can be taken by surprise. This would work well only if left-handers are a small minority, like the 10 percent they are. But it's important to know if left-handers actually win more often in these sports, and only one study—of the World Wrestling Championships—confirmed that. Also wars are no longer fought hand to hand, so the fighting hypothesis should only weaken with time.

So we are left with all kinds of questions, only few definitive answers, and a central fact that can't be argued: 90 percent of us are right-handed.

 **Did You Know . . .** Five out of the last eight American presidents have been left-handed (Trump is right-handed). The only southpaw Canadian prime minister seems to have been Joe Clark (at least in the last fifty years).

Is it true that we use only 10 percent of our brain?

EVEN THOUGH ONE ESTIMATE suggested that fully two-thirds of the public and *half* of science teachers believe that we use only a tenth of our brain, this is simply not true. It doesn't even make sense. Our brains are energy hogs, using up to 20 percent of the energy we take in. Why maintain a brain with that kind of appetite if 90 percent of it sits idle?

Neuroscientists know enough about the human brain now to be able to explain what many of the neural circuits do and how different each is from the next. Memory involves an area close to the ears called the hippocampus; visual information is processed at the back of the brain; planning for the future is located in areas toward the front. It's a complex layout, with activity all across the gray matter.

A catalogue of head injuries over the centuries reveals that any area of the brain that dies or is destroyed leaves the victim compromised. Significant damage causes the loss of a physical or mental ability.

That doesn't mean that the entire brain, all 86 billion neurons (plus many billions of other kinds of cells), is active every second. If your eyes are closed, your visual areas will be less busy than when they're open. If you're not speaking, some circuits in the language areas will be relatively quiet. But there isn't any part of the brain that is sitting idle, waiting to be called into action.

Here's an example. Several years ago neurosurgeons in California were recording electrical activity from single neurons in a patient's brain. (And by the way, no electrical stimulation like this has ever revealed completely "quiet" or inactive parts of the brain.) They discovered to their astonishment that one neuron fired only in response to pictures of *Friends*'s Jennifer Aniston. If the patient was shown photos of Aniston, no matter what the angle or lighting, that faithful neuron fired. And it ignored Kobe Bryant, Julia Roberts, Oprah Winfrey, and Pamela Anderson, although it's likely they all had their own neurons somewhere.

On further investigation, it turned out that this neuron didn't specialize in Jennifer Aniston. It fired in response to "seeing" Lisa Kudrow, too, one of Aniston's fellow characters on *Friends*. And most scientists suspect that this isn't the only neuron that responds to the two of them. There might even be thousands, and given the trickiness of recording from single neurons, it's no surprise that the others haven't been located. And finally, this neuron isn't the only one that responds to famous people: another fired when looking at Luke Skywalker, but like the Aniston neuron, played the field and responded to Yoda, too.

If I only had a brain.

Before you read that last paragraph you likely weren't thinking of *Friends* or Jennifer Aniston or Lisa Kudrow. But now you are, and so the neurons that were sitting there, just being neurons, are now firing rapidly as you see those two actors in your mind's eye. Is it fair to say those neurons represented some tiny percentage of your brain that you weren't using? Maybe. But "weren't using" and "never using" are very different. These neurons haven't fallen into disuse—they're assigned to *Friends* actors (and maybe much more)!

So where did this well-known saying come from? Some scientists, especially the late Barry Beyerstein at Simon Fraser University, have invested a lot of time tracking this down, and I'm convinced Beyerstein has found where it all started.

It seems to have begun with one of the most celebrated psychologists of the late nineteenth and early twentieth centuries: American William James. He was smart and enormously influential. In a lecture to the American Philosophical Association in December 1906, James said, "We are making use of only a small part of our possible mental and physical resources."

In the rest of the lecture he made it clear that he was talking not about a largely silent brain, but about myriad factors, like fatigue and even social convention, that prevent us from thinking freely and energetically all the time.

Somehow James's claim that we don't stretch ourselves to use all our mental resources was transformed into a percentage of brain tissue that we just don't use—not at all what James meant. In the 1920s and '30s, the 10 percent idea became widely quoted. The most influential apostle of the idea was Dale Carnegie, author of *How to Win Friends and Influence People*, one of the bestselling books of all time.

As wrong as it is, the 10 percent idea has persisted for probably a century. Why? Part of it is that we all like to think we have more potential than we've realized. The self-help industry is founded on that. It's also true that believers in strange and unproven phenomena, like ESP, can point to the unused 90 percent and claim that's where it all could happen, if only you could train yourself to use it.

Regardless, it's still true that you can learn a seemingly unlimited number of new pop song lyrics, but while you're doing that, you're forgetting other things. Sad but true: we do use *all* the brain.

A very strange sidelight to this question is that some people have much less brain tissue than the rest of us and yet seem to live pretty normal lives. The most controversial version of this story dates back to 1980, when a British pediatrician, John Lorber, reported he had seen a college student with an IQ of 126 (average is 99) who had almost no brain. But Lorber never published an account of this or other of his subjects in the scientific literature. Doubts were raised about the quality of the brain images Lorber made, and his claim faded away.

However, more recent cases suggest there might be a kernel of truth to this. In 2007 a forty-four-year-old man in Marseille, France, went to a hospital complaining of weakness in his left leg. The medical examination revealed that he had a condition known as hydrocephalus, where fluid fails to drain properly from the brain. Despite the periodic insertion of a shunt to improve the drainage, excess fluid and the internal pressure it caused had pushed against the man's brain, resulting in a huge open space inside the skull, filled with fluid rather than brain tissue.

The MRI images were actually shocking. Doctors estimated that he had lost 50 to 75 percent of his brain. What was left was confined to a thin layer of tissue pressed up against the inside of his skull. That might persuade anyone that this man was barely alive, let alone functioning. Yet his IQ was 75. That *is* below average, yet he worked for the government in Marseille and had a family. A clearer image of exactly how his brain works, how the existing neurons are arranged and communicate with each other, will be known only if an autopsy is done when the man dies, which might not be for several more decades.

What is the explanation for cases like these? Some experts argue that brain plasticity, the ability of the brain to cope with sometimes serious damage, is at the heart of it, but you'd think that at some point the amount of brain tissue would be important.

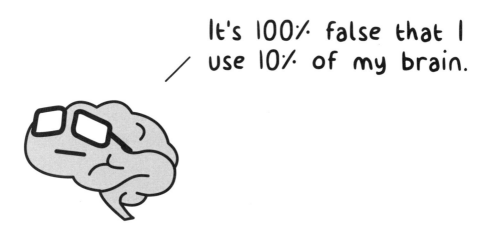

It's 100% false that I use 10% of my brain.

Why do paper cuts and stubbed toes hurt differently?

Not only do paper cuts and stubbed toes hurt *A LOT*, each has a unique feel: the paper cut, sharp and acute; the stubbed toe, a shattering pain followed by one long throb.

These two common accidents illustrate just how different various kinds of pain can be—and they also illustrate some of the complexity of our suffering.

The nerve cells that start both terrible experiences are called nociceptors. Scattered abundantly throughout the skin, they remain inactive until they sense any significant mechanical, chemical, or temperature anomaly. Then they fire. There are two principal kinds of nociceptors: slow reacting and fast reacting. Both transfer their message over to the spinal cord; it then ascends to the brain. The slow C fibers move impulses along at about a meter or two per second, because they're hampered by the lack of an

insulating sheath of myelin around them. The fast A delta fibers have a thin myelin sheath and so can reach ten times that speed.

Science _Fact!_ _Our toes and fingers are especially well endowed with nociceptors—pain detection cells. That is not surprising: we move on our feet and manipulate things with our hands. We can't afford to have them damaged. In fact, in our hunter-gatherer days, if we broke a toe and were unable to walk, hunt, or escape our predators, the injury could prove deadly._

Crushing your baby toe against the fireplace ledge creates enough extreme physical pressure to excite the mechanical nociceptors, which then pass on the message to the spinal cord. Once your toe has been stubbed, there's not much you can do in response (except see Did You Know?). In other situations, where action is required, like jerking your hand back from a flame, your brain is able to execute that move because the fast fibers have relayed the message so quickly. The agonizing, lingering pain or ache that follows is mediated by the slow fibers.

One of the important factors in how much pain you feel with any injury is the number of receptors reporting. Your baby toe is not well protected by muscle or fat, and so the blow is felt over a fairly wide area—the number of activated nociceptors is huge. It's also true that when you hit your baby toe, your entire leg is swinging forward with force, and that force is concentrated on this toe. So more nociceptors reach their threshold for firing.

In contrast, it's not completely understood why paper cuts hurt so much. Mechanical nociceptors are responsible for responding to cuts as well as impacts. Deep cuts, like a surgeon's scalpel slicing through your skin (so-called incisional cuts), induce a complex set of reactions involving these nociceptors. But it's not clear that the paper cut, which is quite superficial, is the same thing.

However, the shallowness of the average paper cut is important. The pain receptors sit in the upper layer of skin, the epidermis, which averages only about a tenth of a millimeter in depth. Paper cuts, although deep enough to trigger the pain sensation, are usually too shallow to bleed, meaning that they won't form a protective scab. The fact that the wound remains open to the air might account for some of the lingering pain.

Another reason that a paper cut, while shallow, hurts so much might be because the edge of the paper, at a micro level, is not smooth, but rough and serrated. (While paper seems so pliable and harmless, when it hits your skin edge on, it's a little stiffer and more dangerous.) It therefore makes a ragged cut. That cut may seem small, but its uneven edges involve more pain receptors than a straight cut.

Science _Fact!_ *When pain signals arrive at the brain, they are sorted, evaluated, and in many cases, dulled. You scrape your knee playing soccer but are unaware of it until after the game. Soldiers, some of them gravely injured, even close to death, may deny they're in pain and refuse painkillers. In the end the amount of pain you feel is determined by the brain, not by the nerve cells reporting it.*

Finally, our brain has a lot to say about how we feel pain. And perhaps unfortunately, large portions of our gray matter are devoted to signals from parts of our bodies like the tongue, lips, fingers, and toes. The end result? The annoyingly common stubbed toe and the slightly less frequent paper cut manage to whip up far more pain than you might expect!

Did You Know . . . One tactic that can reduce the sensation of pain is to rub, squeeze, or otherwise apply new touch sensations to the wounded area. That adds dramatically to the amount of information flowing to the brain, diluting the focus on pain. But this approach probably won't help with stubbed toes or a paper cut. (Rub a paper cut? Are you kidding?) However, many of us employ another coping mechanism that might just do the trick—swearing.

In an article published in the journal *NeuroReport*, Richard Stephens, John Atkins, and Andrew Kingston suggest that cursing helps us deal with pain. The experiment was straightforward: Students were asked to list five words that they might say if they hit their thumb with a hammer. Then they were instructed to utter the first word on the list as much as they wanted as their hand was put into freezing water, leaving their hand submerged as long as possible. A control group was asked to list five words to describe a table and then use one of those while their hands were in the water. So while one group were yelling "#%&*^@@+," the others were screaming "dining" or "pool" or "card"!

The swearers lasted longer in the icy water, had a higher pain threshold, and increased their heart rate. It works! The researchers speculated as to some possible mechanisms, but the effect remains somewhat mysterious.

One disappointing aspect: socially acceptable swear words (yes, there are some) don't work as well. You have to really let loose.

Is gray hair really a thing?

OF COURSE, YOU SAY! *There's a multibillion-dollar hair-dye industry devoted to covering it up.*

But is gray hair really gray or simply a mix of dark and white hair that only *looks* gray from a distance? In other words, could people with full heads of so-called gray hair be able to pull white and dark hairs from their head at the same time?

I thought this might actually be the case, but I was wrong: the hairs on your head do gradually lose color, becoming grayish on their way to ending up totally white.

I'm pretty sure I'm going gray.

Here's how this works. Hair is extruded from deep inside a narrow tube in the skin—the hair follicle. This follicle is actually a tiny organ (with about twenty different kinds of cells); at its base are the materials that create the shaft of the hair, including the keratin protein that gives it strength and the melanocytes that give it color. The melanocytes introduce melanin or pigment granules into the proteins of the hair shaft, and the hair slowly pushes its way out. As the hair makes its way to the surface, sebaceous glands add oil, but the living cells making up the shaft die. The hair that you see is dead—which is why you don't need a painkiller for a haircut!

The hairs on your head grow about a centimeter (half an inch) per month and continue to grow for up to ten years, after which the strand falls out and the follicle begins producing a new hair.

While follicles go through this cycle ten or more times, they produce colored hair only on young heads, as eventually the pigment cells, the melanocytes, just can't keep up. It's not completely clear why. Likely, the aging process reduces either their numbers or their effectiveness. Whatever the reason, eventually gray becomes white as all the melanocytes die off, leaving no more pigment to color the hair.

Science _Fact!_ The 50:50:50 rule states that by the age of fifty, 50 percent of people will be 50 percent gray.

It's a gradual process, at least in most people, but for centuries there have been sensational stories of hair turning white literally overnight. Marie Antoinette, queen of France, was sentenced to death by guillotine for high treason in the French Revolution. Numerous reports claimed that her hair turned white the night before her execution: according to the memoirs of her lady-in-waiting, "sorrow had blanched the Queen's once beautiful hair."

Sir Thomas More apparently had the Antoinette reaction to his own death sentence for treason, too. Annie Oakley, a famous sharpshooter in Buffalo Bill's Wild West show, was injured in a train wreck and claimed that her hair turned completely white over the next seventeen hours. There are many more: King Henry III of Navarre, Mary Queen of Scots, the Mughal emperor (and builder of the Taj Mahal) Shah Jahan, and a long list of less-well-known individuals whose hair, beard, or mustache turned white overnight.

Going gray is the least of my worries.

While these stories are impossible to verify, one intriguing study on "overnight" white hair was done by the nineteenth-century neurologist Charles-Édouard Brown-Séquard, whose own beard turned white in the strangest of ways. When he was forty-five, he found some white hairs in his beard and plucked them. A few days later, more white. He plucked those, too. He did this over and over for a month and a half, and claimed there was no doubt that the new white whiskers that he found every few days were completely white along their entire length and had apparently become so overnight.

So what is going on here?

Turning white overnight would seem impossible. Since hair grows only about a centimeter a month, its overnight new growth would be hard to notice. How then could an entire hair, which a day before was completely colored by pigment, turn white? Where would that pigment go?

There have been suggestions that somehow air gets into the pigment granules, which would indeed make them look white, but there's not much evidence for that.

There is one scientific explanation that probably accounts for some of the cases. Alopecia areata is thought to be an autoimmune disease, in which the body's immune system mistakenly attacks its own tissues—in this case, the hair follicles. This causes hair to fall out, sometimes in small patches, sometimes all over. Most often the hair grows back eventually, although some of it might come back in white. But that's a process that takes weeks, not twenty-four hours.

If, however, a person already had some gray hair and patches of pigmented hair fell out, that could leave behind only the white and gray. At first glance it would appear that the entire head of hair had turned white—instantly. As you can imagine, though, this would have to happen uniformly all over the head to be convincing.

Science Fact! *Scientists have plotted how the graying process spreads across our heads. It starts at the temples, then spreads upward to the top of the head and slowly to the back.*

In the end, even though there are published claims of "authenticated" cases of hair turning white overnight, there are few good explanations of how that might have occurred, making most scientists skeptical that these ever happened.

But if they did, there may be a very simple explanation for a few of them: a lack of hair dye.

Some have argued that Marie Antoinette had white hair at least several months before her execution. Would it be any surprise if her jailers didn't let her turn her cell into a hair salon?

Do human beings molt?

THE ANSWER IS YES! But in case you're thinking this might be the perfect weight loss technique, think again, because for every skin flake that drifts off into the air (adding to the dust in your house or the air in the subway), more are being created below the surface of your skin. The skin is essentially a conveyor belt, generating new cells at the bottom and getting rid of them at the top. If you think about that sequence, you will understand how it's possible that every person loses an incredible 35 kilograms (just over 77 pounds) of skin over their lifetime.

The thickness of our skin, or epidermis, differs across the surface of the body, from as little as 0.05 millimeters (a little less than one six-hundredth of an inch) on our eyelids to 1.5 millimeters (one sixteenth of an inch) on the palms of our hands or the soles of our feet. The physical stress experienced by hands and feet requires a thicker layer.

You think *you* molt?

Skin is a multilayered organ with the job of preventing water from escaping and bacteria from invading our bodies. But because it is frequently struck, cut, pinched, or scraped, the skin must constantly renew itself to maintain its integrity. So while, unlike other organs such as the lungs or heart, it may seem completely static, it is actually seething with activity.

Science _Fact!_ _Reptiles like snakes and lizards shed their skins, too, but not one flake at a time. They build an entire new layer below the surface, then molt, sometimes losing the entire outer layer in one piece._

The basement layer of our skin, the innermost part of the epidermis, is furnished with stem cells: living cells that have not yet committed to becoming any sort of specialized skin cell. Their potential to turn themselves into a variety of cells is crucial to a healthy skin, because sometimes an injury requires an unusually large number of specific kinds of cells and stem cells are responsible for supplying them. As these cells at the base multiply, they must migrate somewhere, and the only option is up toward the surface.

As they move upward, they change by shutting down the synthesis of some components and ramping up others. The result is that what were originally round or oblong cells begin to flatten as they move upward.

Their final form is a dead, flattened cell that has lost most of its internal components, like the nucleus, and is instead filled with a network of the protein keratin—the same stuff that makes up your nails and hair. These dead cells lie fifteen to twenty deep, held tightly together by chemical bonds, or cross-links, and embedded in a film of fatty material, including cholesterol, to produce the top layer of our skin, called the stratum corneum. (This is only a small percentage of the entire thickness of the skin.)

Gradually, the cross-links between these surface cells deteriorate. When the final links have been severed, the dead cells are free to break away and leave the surface, and do so in astonishing numbers: 500 million a day is the best estimate. Their departure, although fantastic in number, has no impact on the skin because they have already been replaced by the next layer of flattened cells, which are already in preparation for separation themselves.

Our entire outer layer of skin gets shed every month or so. Estimates vary, but it takes roughly the same amount of time for a cell generated in the lowest layer to make it to the surface.

Did You Know . . . Where do all the dead skin cells end up? As I mentioned earlier, they collect in the dust in your house as well as in the air around you (especially in semi-enclosed spaces like subway tunnels). But of course you're shedding skin flakes twenty-four hours a day, meaning that a lot of them end up on your bedsheets and in the mattress. Fortunately (unfortunately?) there are dust mites, tiny animals distantly related to spiders, living in your mattress, consuming your skin flakes. And mating. And laying eggs. So no worries: a substantial number of flakes will be consumed right where you deposited them.

Now let's do the math to support the claim that we lose 35 kilograms of worn-out skin cells every year. Number shed every day: 500 million. Their individual weight is infinitesimal, but the numbers make up for that, meaning that we shed somewhere between 30 and 90 milligrams worth *every hour*. That's roughly 1,400 milligrams per day, a gram and a half or the equivalent of a third of a teaspoon of sugar. And yes, when you multiply that by 365, it amounts to slightly more than 35 kilograms every year (but 35 is a nice round number). And imagine this: if you live eighty years, even allowing for fewer skin cells shed when you were a child (because you had less skin!), you've probably contributed about two metric tons (2,200 pounds) of skin to the world around you. How generous!

Did You Know . . . With the conveyor belt of skin cells churning out new flakes every day, why do tattoos remain? Because they're deposited below the layer that's active: in effect they're steps away from the conveyor and so don't get swept up by it. Tattoos can fade with time, partially by exposure to the sun, but unless they're lasered to break up the tiny drops of ink, they're in place to stay.

Have we peaked as a species?

It's been a long, hard road to get to where we, *Homo sapiens*, are today. A few million years ago our ancestors barely walked upright, had small brains, and probably spent a fair amount of time in the trees. Even a million years ago, our species didn't yet exist, although our predecessors at the time looked far more human than they did apelike.

Then somewhere around 300,000 years ago, probably in multiple locations in Africa, the first members of *Homo sapiens* appeared. They still didn't look quite like modern people, with their elongated skulls and oversized teeth, but they were the same as us in most ways. Three hundred millennia later, we like to think that we have never been better, and that the next millennia will see even more improvements. After all, why would evolution stop?

But wait. In 2017 a team of French scientists pulled together a huge amount of data to show how far we have come, and they arrived at a startling conclusion. *We have reached our peak.* No matter what you look at—speed, strength, height, life expectancy—the numbers are all leveling off after years of climbing. That's the headline story, but it's worth a closer look at the data.

Life expectancy is a good place to start. When you're born, you and all the other babies born in that year are assigned an age that you're likely to achieve. So, for instance, most North American males born in the year 2000 can expect to live to seventy-seven; females to eighty-two (women always tend to live longer). A hundred years earlier, those numbers would have been at least twenty years less, probably fifty-seven for men, and fifty-nine or sixty for women. That number has been rising steadily for much more than a hundred years. Every four years that pass, life expectancy goes up at least one year, but that number doesn't set a limit on your age. It's just a measure of how long a baby born that year is expected to survive. Some die early; some live much longer. Some scientists have argued that half the babies born in the developed countries in 2000 will live to a hundred.

But can this continue? So far the oldest human ever recorded was Jeanne Calment, who died in 1997 at the remarkable age of a hundred and twenty-two. No one has reached that age since. So that might be our maximum age, somewhere around a hundred and twenty. There are those who think as we understand aging better, we'll push past any such limits, but they're in the minority.

Did You Know . . . Every organ in the human body has what scientists call its maximum functional performance—its performance limit. For instance, the fastest a human heart can beat is just over 200 beats per minute (average is anywhere from 40 to 100). But achieving that number becomes harder and harder with age. In fact when we are in our *twenties,* we are at peak performance. The rest of life is a gradual decline from that high point.

Height is another way of measuring how close we are to a peak. Even though the vast majority of humans fall into a narrow range of height, the total range is amazing. The tallest man ever recorded in the world was Robert Wadlow, who stood just shy of nine feet tall (272 centimeters) when he died at the age of twenty-two. He was five times taller than the shortest man, Chandra Bahadur Dangi, who was 55 centimeters (just under two feet).

There's no doubt we are taller today than we were a few centuries ago. But the idea that somehow we've been gaining height steadily for centuries isn't true, no matter how many tiny suits of armor are in your local museum.

Here's a time line. One of the earliest species in our genus, *Homo erectus*, represented a leap in height from earlier ancestors, and likely stood somewhere between five and six feet tall. And that was a million years ago. For millennia after that, evidence is scanty, but height may have remained more or less constant until about 18,000 years ago, when it started to decline dramatically. From 18,000 to 1,000 years ago, men lost an average of

Tall **Small**

10 centimeters or 4 inches. The move to agriculture might have been responsible, because inconsistent crops and more crowded living could have negatively impacted health.

Then from the 900s to the 1300s we see average heights in England (derived from skeletal remains) at five-seven for men, five-two for women—not very much different from today. But things changed dramatically in the 1600s and 1700s, where average heights declined to their lowest point in the records—two and a half inches shorter than they'd been centuries earlier. (Yet at this time the indigenous people living on the North American plains were in the five-eight to five-ten range!)

Did You Know . . . King Henry VIII, who ruled at the beginning of the 1500s, was over six feet tall and likely impressive for that, as his average countryman stood only about five-five.

Every inch a king.

Since that time there's been a remarkable global growth in height, more than three inches on average, maybe as much as four inches. Today, Dutch men (six feet) and Latvian women (five-seven) are the tallest in the world. But these rapid gains over the last century or two seem to be slowing, even leveling off.

French researchers also turned to sports data to see if we have reached our peak. After all, world-class athletes represent the extreme of our human capabilities. And once again, it looks as if we may be at our apex.

It may not look like that at first glance: throughout the entire twentieth century, records in weight lifting, running, jumping, swimming, skating, skiing, and cycling were set steadily, the only exception being during the two world wars.

The men's world record time for the marathon has fallen from 2:55:18 in 1909 to 2:02:57 in 2014. The hundred-meter freestyle in the pool has dropped from 65 seconds in 1905 to 46.91 seconds today.

There are innumerable examples like that, but researchers argue, with hard numbers behind them, that for the past thirty years, the rate of change of world records has been slowing for both men and women: the records aren't falling as fast or as far. This is in spite of abundant drug use in the last few years that has artificially raised performance.

This squares with some predictions, based on human muscle mass, heart rate, oxygen uptake, and stride length, that we are getting close to maximum performance. The sad truth? This is probably as good as it gets.

I'm always at
peak performance.

Why are fingernails on a chalkboard so obnoxious?

Even though chalkboards and chalk are gradually being replaced by whiteboards and markers, most people have still heard—and suffered from—the sharp, grating, penetrating sounds of fingernails being scraped across the board. The sound is cringe-making. Just perching your fingernails on a chalkboard is enough to make most people squirm. But why is it so disturbing?

There is no single answer to this question, but several suggestions that, when taken together, hint at what might be going on.

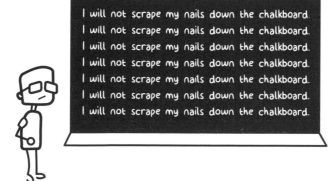

First, it goes without saying that the chalkboard sound is more noise than music. Noise and music are made in different ways. Think of a string on a guitar: when it's plucked, it vibrates, those vibrations are felt through the entire instrument, and the sound that results is a mix of different frequencies of vibration. The frequencies created depend on the thickness and tension on the string, the movement of the surface of the instrument, and the air inside. Some frequencies are enhanced, some diminished, and sound waves are created in the air surrounding the guitar. That is what you hear.

The different frequencies produced by the guitar string are made up of the fundamental ones and then a set of overtones, or higher frequency harmonics; all of them together create the characteristic sound of a guitar. The same note on a piano or a trumpet sounds different because different harmonics are emphasized. A pure tone, one with no harmonics, can be generated, too—there are websites that will do it for you—but the absence of harmonics makes the pure tone pretty uninteresting as a result. All these harmonics have a mathematical relationship to one another and in general sound pleasant to us.

Noise is quite different from music. Instead of a fundamental tone and its harmonics, noise contains just about every frequency, some louder than others. It's a mishmash of sound. Think about what happens when you scrape your nails across a blackboard. It's not a smooth process; your nails stick, then slip, then *catch*, then stick again. All those movements cause them to generate a mess of sound waves ranging from very low pitched to very high, so the whole effect is pretty dramatic.

However, this fundamental difference between music and noise doesn't really explain why this particular noise is so unpleasant (there hasn't been an experiment yet where anyone claimed it was enjoyable). But there are theories.

In the mid-1980s, Lynn Halpern and her colleagues argued that if you scan the set of frequencies produced by nails on chalkboard, they bear an uncanny resemblance to alarm calls made by some primates.

In a paper titled "Psychoacoustics of a Chilling Sound," they described generating a large set of sounds and having volunteers judge their pleasantness—or not. The most unpleasant was a three-prong garden tool being dragged across a piece of slate, a sound that they described as "disturbingly similar" to fingernails on a chalkboard. The scientists were startled to see that, contrary to the common belief, high frequencies were not the crucial contributors to the unpleasantness. (They determined this by filtering out high frequencies and asking volunteers to judge the unpleasantness of the reduced sound versus the original.) It may seem like much of the chalkboard noise is squeaking, but while squeaks are mostly high-pitched sounds, in this case they seem not to matter.

However, sounds in the midrange of human hearing seemed to play a crucial role. The authors went on to speculate that their most unpleasant sound was similar to recorded warning calls of macaque monkeys: harsh, discordant, and sure to attract attention. It does seem like a faint and unlikely evolutionary trail from monkeys to humans, warning calls to chalkboards, but this general idea has been echoed by other researchers.

Marc Hauser and Josh McDermott took a stab at the idea with an experiment designed principally to see if monkeys—they used cotton-top tamarins—responded to pleasant versus unpleasant

sounds. The scientists were interested in whether music meant anything to these monkeys, but as part of the experimental design they included a "fingernails on the chalkboard" sound. The animals were placed in a V-shaped maze and so could stay in one arm and listen to the sound being played there, or move to the other arm for a different sound. The monkeys apparently had no musical judgment—they reacted the same to both harmonious (musical) sounds and dissonant sounds—but they also didn't discriminate between fingernails on a board and white noise, that rushing sound created from a mix of every frequency.

So here are monkeys that hear nothing special about a sound that is supposed to be like the alarm calls of monkeys. But you could argue that cotton-top tamarins and macaques are very different and probably split from each other 40 million years ago, evolving separately since then, so there's been plenty of time to differentiate enough that the call of one means nothing to the other.

Such beautiful music.

Agreed.

The most recent experiments to determine just why fingernails on a chalkboard are so consistently irritating managed to define the frequency range of the sound much more precisely and offer an explanation based on that. Michael Oehler and Christoph Reuter used the usual setup—playing sounds for people and having them judge the sounds' unpleasantness—but were able to tinker with the sounds more than others had done before them. They identified the crucial frequency range as between 2000 and 4000 hertz, right smack in the middle of the human speech range. If those frequencies were diminished, the sound became much less unpleasant. Reuter and Oehler also showed that an unpleasant sound causes the body to react physically, so our judgments of them are more than just opinions.

And while Oehler and Reuter didn't test the monkey call hypothesis, they did point out that the frequencies they found to be most disturbing, those from 2000 to 4000 hertz, are exactly the frequencies that the human ear canal amplifies by virtue of its shape.

That's where the analysis stands today: that dreadful sound might hearken back to our distant evolutionary past (but there's not much evidence for that); it's not the whole sound but a selective part of it that generates the most discomfort; and unfortunately our hearing apparatus is perfectly designed to make it all worse.

Do spouses tend to look like each other?

THE IDEA THAT PARTNERS MAY LOOK STRANGELY SIMILAR may seem odd, but there is evidence that suggests this happens more than you might think. And it's not just something as minor as hair color. One study measured several body parts, ranging from the length of fingers to the size of the earlobe, and concluded that husbands and wives had more physical features in common than did unmarried but similar individuals.

There are a number of theories about why this may be the case. One study presented experimental subjects with images of an unknown person paired with an image of the subject's own face digitally melded with another. Such "hybrid" faces preserve some similarity but not enough to allow recognition. In this case, subjects preferred faces that were partly their own. If this held true, they would likely be attracted to someone who looked like them.

There have also been some commonsense explanations. After all, people tend not to migrate all that much, so many places will be populated by significant numbers of people who are the third or fourth or even tenth generation of their family. Without many arrivals or departures, the gene pool stays pretty constant, and, after all, genes determine looks. In these cases, it isn't surprising that spouses resemble each other.

Did You Know . . . While looking like your spouse might have some evolutionary purpose, looking like your dog seems pointless. Yet, oddly enough, there's science that claims that dog owners do tend to look like their dogs.

One study by Stanley Coren at the University of British Columbia showed women images of four kinds of dogs (an English springer spaniel, a Siberian husky, a beagle, and a basenji) and concluded that those women with long hair preferred the dogs with so-called lop ears that hang down beside their heads, whereas the women with short hair tended to prefer dogs with pricked ears that stand up straight.

If it were just that one study I don't think much could be made of this, but there was another. Michael Roy and Nicholas Christenfeld of the University of California at San Diego theorized that if people buy dogs that look like them, they'd pretty much be forced to buy purebreds, because the adult appearance of such dogs can be safely predicted. It's much harder to predict what a mixed-breed puppy is going to look like. Volunteer judges viewed a series of photos of fifteen dogs and one owner—in each experiment, only one of the dogs actually belonged to the owner. Of twenty mixed-breeds, only seven were matched with the right owner, but there were an amazing sixteen matches out of twenty-five for the purebreds.

Mysteriously though, none of the obvious characteristics like hairiness, size, and sharpness of features seemed responsible for the correlation.

But in 1987 the whole idea took a curious twist when University of Michigan psychologists found that couples look more and more like each other *as time passes*. As the researchers said when they reported it, there was really no reason to think they should, and yet they did.

People say we look alike.

I don't see it.

The couples weren't very diverse: Caucasians from fifty to sixty, either from Detroit or rural Wisconsin. Not representative of the general public, but consistent. Volunteer students were shown the faces of twelve married couples, taken at two different times: at their marriage and then twenty-five years later. But the faces were mixed up so that students had to compare a single face, man or woman, to a possible set of six mates and choose the pair that looked most like each other. No student saw the same couple at both ages and the photos were trimmed to ensure that things like clothing, which could be similar between mates, wouldn't influence the students' assessments.

The students had a much higher rate of success matching the married couples using the photos taken after a quarter century of marriage. (The other strange result was that at the time they were married, the spouses didn't seem particularly similar, as they should have.) The evidence of growing similarity through time came out of the blue, but the possible explanations were even more interesting.

Previous researchers had suggested that diet might be important. Couples pretty much eat the same food, and with time, some might put on weight while others stay slim. That would be manifest as a chubby face or a slim one. But the experimenters here ranked husbands and wives from heaviest to lightest, and they didn't match. So diet, at least from the weight point of view, wasn't the reason. They also eliminated the possibility that climate (sunshine and wrinkles) might play a role. Instead, the team reached back to an idea first hinted at in 1907. Emotions, blood flow, and facial muscles are all interconnected. Over time, if a couple shares emotions and their accompanying expressions, muscles and blood flow could conspire to cause their faces to adopt similar shapes. Those emotions could be positive, negative, or somewhere in between, but the key thing is that they be shared over and over by spouses.

This was a study rich in ideas. It also showed that those couples who looked most like each other claimed to be happier than the rest (possibly because they were more empathetic).

Since 1987, further studies have cemented the sturdy relationship between personality type (as read in a face) and similarity among couples. When we judge a couple to resemble each other, we're not just superficially assessing facial features; we're actually making (intelligent) guesses about personality type and comparing the two people that way as well. If their personalities seem similar, their faces will strike us as more similar. Just another example of the importance we attach to socializing and reading others' faces.

And the argument makes sense. If people are attracted to others who look like them and therefore have matching personalities as well, they might stay together longer and have more children. That's how we got here as a species.

Why do we blush?

I THINK EVERYONE HAS BLUSHED at one time or another. Those with pale complexions might turn a startling shade of pink. Those with darker skin might simply feel an uncomfortable wave of heat ripple across their cheeks and forehead. But while many of us are well aware of the sensations that accompany blushing, why does it happen?

Most of us understand what causes us to blush. It usually happens when we are feeling awkward or embarrassed. Sometimes the blushing itself makes us feel even more awkward and embarrassed—we can feel it happening. But what's the point, exactly?

The answer is fairly straightforward. Blushing is a social signal that indicates to others that we are uncomfortable. And it seems to work. A number of studies have shown that blushing (especially when it's obvious) does convince others that we are regretful—more than words alone do. (Just

You are the apple of my eye.

as well there's some benefit, because the blusher has no control over it, and it often happens when people *least* want to draw attention to themselves.) This also explains why the heightened flow of blood occurs mostly in the face and neck—it wouldn't be an effective social signal if our shoulders or calves turned red! But the story goes much deeper than this; the fact that embarrassment can cause increased blood flow to the face is just the tip of the iceberg.

In early 2018, scientists at Ohio State University published a remarkable study showing that blushing is just one example of facial color changes caused by emotions, and an extreme one at that. It has always been thought that the face conveyed emotions like sadness, anger, happiness, surprise, and fear by the movement of muscles.

Widened eyes or a downturned mouth might be obvious examples, but even much subtler muscle movements betray what we are feeling. The Ohio State scientists wondered if the intricate network of blood vessels under the skin of the face might also play a role.

They acknowledged that a few other strong emotions affected blood flow to the face, and therefore color changes in pale complexions: anger producing red faces (increased blood flow, but not the same mechanism as blushing) and fear leading to extreme paleness (less blood flow). But they searched for the not so obvious.

They began with analyzing nearly two hundred images of people of various complexions conveying eighteen different emotions by their facial expression (in other words, expressions created by the face muscles). They then analyzed those faces for minute color differences.

Now, before we get to their findings, let's address that idea of face "colors." You might not have thought that there was anything more than the essential tone of your skin. But before the Ohio State researchers had started their studies, Mark Changizi, a theoretical neuroscientist at 2AI Labs in the United States, had observed that changes in the amount of blood flowing through the vessels underlying our facial skin, and the amount of oxygen in that blood, can shade the color of our face from reddish to bluish, or yellowish to greenish.

(For instance, press hard on the palm of your hand, squeezing the blood out, and you will see some of the rosy undertones of your skin disappear and your flesh take on a more yellowish hue.) In your face, expanding and contracting blood vessels do the same thing, and in very subtle combinations, these colors create a wide variety of color patterns across the face. Those patterns are driven by the emotion you're feeling.

Did You Know . . . Some unfortunate people blush constantly. Once aware that they are blushing, they can't stop because they are embarrassed by the blushing itself. They may also blush just from the fear that they might blush. No wonder some go to the lengths of surgery to stop this cycle. Blushing can be treated surgically by cutting the nerves that control blood flow to the face. It's called a bilateral endoscopic thoracic sympathectomy. For many people—up to 90 percent—it changes their life for the better, but there are possible side effects, including an increase in sweating and even drooping eyelids.

When the Ohio State researchers ran their two hundred photographs through a computer program, they found that it could identify the emotion being expressed by these kinds of subtle color changes alone. The colors were consistent for each emotion but different for each one. They then tested the idea with faces that were neutral, expressing no identifiable emotion by expression. By subtly adding color to these faces (colors they had already identified as being associated with particular emotions), they showed that volunteers could identify the emotion of a neutral face just by the faint difference in color.

The authors concluded that this might explain why we, of all primates, have no hair on our faces.

Did You Know . . . Charles Darwin called blushing "the most peculiar and the most human of all expressions."

Mark Changizi's earlier work supports this idea. He has argued that the commonly held belief that color vision in humans—and other primates—evolved so we could identify ripe fruit and fresh leaves by color is misguided. (Other mammals, non-fruit eaters, can't see the red-green-yellow part of the spectrum.) Changizi points out that there's great variation in the colors of fruit eaten by primates around the world, yet only a single kind of color vision. Changizi suggests that it is more likely that our color vision developed so that we could better read emotions as they played out on our skins.

Did You Know . . . Mark Changizi argues that despite the fact that the face is a beacon of emotion, we are practically blind when it comes to labeling its color. An analysis of languages around the world showed that there are many colors common to all languages, but none of them actually describe facial color. Is your face pink? Brown? Chartreuse? Whatever color I suggest, you'd probably say it's inaccurate. Changizi thinks that's because we need to be able to tell minute color differences from face to face, and even within the same face from moment to moment. By not having a stereotypical idea of face color, we are free to be able to discriminate among those tiny changes.

If you ask anyone if they can recognize emotions in a face, they'll say yes—they know the difference between a frown and a smile. But in truth they're much better at it than that; they can also see, although they don't know it, minute color changes that convey the same information.

Science is cool.

Science is hot.

History Mystery

Did Shakespeare really write all of his plays?

WHAT? THAT QUESTION MAKES NO SENSE. HIS NAME IS ON THEM!

Maybe so, but there have been doubts ever since Shakespeare's own time about whether one of the greatest writers in the English language actually wrote his own stuff. And while literary experts and historians have marshaled arguments on both sides of the question, it may be the scientists who have the final answer.

The question about Shakespeare's authorship really falls into two parts. One, did he write every word of every play? And two, did Will Shakespeare of Stratford actually write any one of those plays? And if not, who did?

But why is Shakespeare doubted at all? The main argument is that his plays show evidence of an incredible

By me, or not by me? That is the question.

breadth of knowledge of courtly life and foreign countries that Shakespeare, a common man, not a courtly one—a man who never left England—simply wouldn't have. This line of reasoning is made easier by the shortage of information about Shakespeare's life. We know his father was a tradesman, he married Anne Hathaway, and that he became an actor and a producer of theater in London. But we know next to nothing about his schooling, and he's not known to have traveled. His signature appears on a handful of documents, but there doesn't exist a single play, or even part of one, in his own writing. And he mentioned no books or manuscripts in his will.

Skeptics have used this lack of biographical detail to argue that Shakespeare is clearly not worldly enough to have written about Italy (*The Merchant of Venice*) and Denmark (*Hamlet*) or about the lives of kings and other historical figures.

Defenders of Shakespeare dismiss this as simply snobbery. Why couldn't a brilliant writer emerge from a relatively nondescript background? Why couldn't he have learned about court and foreign countries from people who had been there or from other written accounts?

The arguments, both pro and con, rest on very little evidence, but the story gets even murkier when it tries to identify who actually *did* write the plays.

The names of at least seventy alternative authors have been thrown into the ring, and the list of those convinced that the plays were written by someone else is pretty impressive, including Mark Twain, Sigmund Freud, Charlie Chaplin, Orson Welles, Helen Keller, and Malcolm X.

Among the seventy contenders, two stand out as the most likely, or at least most popular. One is the playwright Christopher Marlowe. Marlowe was an excellent playwright, and his career and Shakespeare's overlapped for a few years. One significant problem with Marlowe's candidacy is that he was stabbed in the eye over an unpaid bill and killed in 1593, long before the majority of Shakespeare's plays were actually written. For those who believe

Marlowe wrote the plays, it's necessary to argue that his death was faked (or the whole story fabricated) to allow him to live—and write—in anonymity. A rumor that he had been a spy supposedly explains his need to lay low, write plays, but use Shakespeare's name as author. Pretty farfetched.

Edward de Vere, the 17th Earl of Oxford and the hero of the 2011 Roland Emmerich film *Anonymous*, is now probably the leading candidate. He ticks off all those boxes that Will Shakespeare misses—like education and familiarity with court life (he graduated from Cambridge at fourteen, did a master's degree, and studied law), travel (not only did he spend months in Italy, he also, intriguingly, was kidnapped by pirates and set ashore naked in Denmark, just as *Hamlet* was "set naked" there)—and as if that wasn't enough, his nickname was "Spear-shaker"; the family coat of arms had a lion wielding a spear.

All of this is tempting, but also seems to contain a fair amount of rumor and speculation.

So this is where science comes in.

The reason that there is so much confusion surrounding the plays is that writing and publishing were so different in Shakespeare's time. There are no original manuscripts of the plays in existence, just published versions that have been interpreted, altered, even rewritten to the satisfaction of whoever was publishing them. A number of different versions of the same play exist dating back to Shakespeare's own time. Screening them for authorship is therefore incredibly difficult.

Nonetheless computer programs can be used to determine whether the writer is favoring certain words or phrases or combinations. For instance, one approach has been to characterize words as "Shakespeare-plus," words that appear more often in Shakespeare's work than in the writing of his contemporaries, and "Shakespeare-minus," words that appear less often. Shakespeare was also tricky, changing nouns to verbs and adjectives to nouns. Writing algorithms using this kind of information showed the three *King Henry VI* plays (Parts I, II, and III) to be more Marlowe than Shakespeare. Scholars have subsequently given Marlowe cowriting credit for these works.

(In fact, a third author, Thomas Nashe, is now believed to have written the first act of *Henry VI*, part 1.)

A different computer program has been used to argue that the Earl of Oxford couldn't possibly have been the writer of the plays. Even if we set aside the fact that he died in 1604, making it as difficult for him to write the several later plays as it was for Marlowe, it's the writing style that is at issue.

A group called the Claremont Shakespeare Clinic took on the issue of the Earl of Oxford versus Shakespeare. Their detailed analysis of the writing in the plays, including word lengths, sentence lengths, and hyphenated words, when compared to the known poems of the Earl of Oxford, show, at least to their satisfaction, that the earl is not the mystery writer. "The odds that either could have written the other's work," they concluded, "are much lower than the odds of getting hit by lightning." No doubt where they stand!

Nevertheless, as a result of these new analytical tools, scholars now give other authors shared writing credit on seventeen out of a total of forty-four Shakespeare plays.

The evidence that Shakespeare may have had a little help does not, however, support the idea that he didn't write any of the plays at all. And it hasn't seemed to dim his star.

In the end, perhaps the testimonies of Shakespeare's contemporaries should be remembered. Many of them expressed their admiration for his talent, like his friend the poet Ben Jonson: "Soul of the age! The applause, delight, the wonder of our stage!"

Maybe that's where all the doubt comes from: a talent so great that it's hard to imagine any one person being responsible for it all.

Part 3
Weird, Wacky, and Wondrous

Why do shoelaces come undone?

THE FIRST SHOES THAT RESEMBLE TODAY'S, with laces and eyelets or hooks, are at least a thousand years old, and four thousand years before that people were using strips of material to help keep shoes on. Shoelaces have been coming undone since then, yet amazingly, the first explanation of why this happens was only offered in 2017.

At least now we know, after centuries of puzzling, why this happens so often. A warning, though: I am not going to try to describe the differences between granny knots and square knots and why you should use square ones if you don't want loose laces. There are videos and explanations on the web for that. I'm sticking to the science of why they come undone.

I'm all tied up in knots

Shoelaces come undone because of the forces applied to them by walking or running.

Even walking, much more sedate than running, is enough to pull those knots apart. We walk in a fairly predictable way, starting with a fall. From a standing start, you start to lean, then fall forward, but at the same time swing a leg out in front of you to break the fall. The foot of that leg strikes the ground heel-first, then rolls forward toward the toes, finally pushing off the big toe. As that's happening, the other leg is already in swing. And that's it.

Did You Know . . . The average adult walks about three meters per second (five miles per hour), a speed limited by the length of our legs. The longer the legs, the less the torso bobs up and down and the faster the walk. The most energy-wise pace of walking is about a hundred steps per minute, but studies have shown that in big cities, both men and women walk faster than that. And while all this is happening, you're spending about three-quarters of your walking time on one leg, more if you're walking faster.

Could there be something more efficient than either walking or running? Scientists Andy Ruina and Manoj Srinivasan wondered if there could be other good gaits that for some reason we don't use. When they asked their computer program to come up with the most efficient ways of moving around on two legs, the computer came back the two you'd expect, running and walking, but also a surprise third, a mix of running and walking they called pendular running. It's a weird gait, something like walking but punctuated by moments when both legs are off the ground, like running. No one anywhere has been seen to use it, so it must have some significant flaw.

University of Berkeley scientists Christopher Daily-Diamond, Christine Gregg, and Oliver O'Reilly set out to understand why shoelaces become untied. But before we get to their discoveries, we need to understand what's holding the knot together. That's the friction between the two laces, especially in the center of the knot. To some extent you can increase it by tying the knot as tightly as you can.

What's tearing it apart? First, every heel strike creates a force somewhere between six to eight times the force of gravity. That force jolts the center of the knot, causing it to open up slightly, and that provides an opportunity for the laces to lose their frictional grip on each other. That's one factor. The second is the swing of the leg, which causes the two free ends of the lace to flop back and forth, accelerating the slippage.

These two factors combine at the critical moment, when the leg, which has been moving forward, suddenly stops as the heel strikes the surface. For a moment then, the free ends of the laces and the loops are still moving forward by their own inertia, but the knot, which isn't free to move, has stopped. And the knot is opening up and the laces are gradually pulled through it, very slowly at first because the center of the knot is still firm and frictional forces are high. But as the knot center continues to loosen and the laces continue to swing, the knot reaches a turning point. The whole thing can come apart in as few as two or three steps, or as the researchers put it, the knot undergoes "gradual loosening and acute failure."

Neither of the two critical factors, the heel strike and the leg swing, is enough on its own. The researchers showed that sitting on the edge of a table and swinging your leg back and forth will not cause shoelaces to become untied, nor will stomping your foot on the floor. And while I vowed I wouldn't try to depict with words the differences between good and bad knots, there's no doubt some are better than others, probably because of the geometry of the knot when tied.

The research team hasn't yet tested different kinds of laces, though they did test the laces they used with different sneakers, hiking boots, and running shoes. So how can you make use of this groundbreaking information? Learn the differences between knots and use ones that resist this collapse. Or don't do the laces up, or don't even put laces in your shoes. Or wear sandals. Just no Velcro.

Does holding my key fob to my head make it work better?

Forgot where you parked your car at the mall? Standing there forlornly, arms full of packages, pressing the button on your key fob over and over, hoping for the sound of the horn? Nothing much can be done about your inattention when you parked, but there is a trick that might help you find your vehicle. Press the key fob against your head!

Why on earth would you do that? Because it will extend the range of your fob, perhaps as much as 30 meters (about 100 feet), and might just allow it to trigger your car horn after all. But how can this possibly work?

I'm feeling all keyed up.

The electronics of a key fob are pretty straightforward (that is, if you think electromagnetic radiation is straightforward). The key fob contains a battery; when you press the button on it, a circuit inside the fob closes, current flows, and finally a transmitter fires up and emits a signal in the radio wave band of the electromagnetic spectrum. Each wave is about a meter long, roughly three feet, and in North American cars they're emitted 315 million times every second. (In other words, their frequency is 315 megahertz. For comparison, FM radio stations use frequencies ranging from 88 to 108 megahertz.)

Those waves spread out from the fob, and if they wash over your car, a receiver tuned to that frequency will be alerted. But most cars parked around yours are sensitive to that same frequency, so what prevents your key fob from unlocking all of them? A code.

Both the fob (the transmitter) and the receiver in your car share a code of eight numbers. But there's a trick involved. Both your key and your car change the code every time it's used by randomly generating a new set of matching numbers. This is known as a rolling code. So even if someone standing near you in the parking lot had the electronic technology to intercept the signal as you lock your car, they couldn't use that same code to open it because a new code would have already been put in place.

The last part of this process is simple. Once the receiver in the car has verified that the signal is from you and no one else, it triggers an action, such as opening the door or locking it or turning on the security alarm. This part is exactly like driving a radio-controlled toy: the radio waves turn on the motors that drive the actions.

 **Did You Know . . .** American Samy Kamcar researches security and, as part of that, invents hacks. He claims to have come up with a way to overcome rolling codes on remote locks. Here's how. You walk away from your car and thumb the fob to lock your car. In a nearby car, a thief intercepts that signal, preventing it from getting to your car. You try again (this time your fob sends a new code) and the thief intercepts that, too, but at the same time releases the first code to the car so it locks. You walk away happily thinking you've locked your car. The thief, meanwhile, still has the second code, which can now be used to open the car. Kamcar points out that it would be easy to foil this by having each code time-limited: it would exist, then vanish.

Because the key fob's radio waves are not laser-focused, they spread out as they travel, and if you are too far away from your car, the amount of radiation the car receives will be under a critical threshold, and nothing will happen. That distance for most cars is about 30 meters, or 100 feet. If you happen to be standing 40 meters away (about 130 feet), you're out of luck!

This is where your head comes in. It has been demonstrated—at least to my satisfaction—that you can turn on the lights and unlock the car from farther away if you're holding the key against your head. But if you're hoping for a clear-cut explanation of why this works, you're probably going to be disappointed.

Skeptics argue that it's probably the height of the fob off the ground, not the fact that it's touching your head. A physicist in the United States, Chris Mueller, did a lengthy set of experiments, varying things like height above the ground, direction of the fob, and the exact placement on the head, and concluded that there was a positive effect of the head, but also that height wasn't important. Holding the fob against your temple was the best, although pointing it at the car also helped. For some reason holding the fob at the chest didn't work.

I am the key to my own success.

I have been involved in an experiment like that and, like Mueller, found that the head was definitely the most effective body part, and that you could take several steps beyond the normal range of your car (a little over 30 meters in our case), try the fob—nothing. Then hold the fob to your head and flash! The lights are on.

So what's going on? One idea is simply that the signal reflects off the inside of the skull and so is reinforced as it sets off toward the car. Alternatively, the skull can act as a resonant cavity and amp up the signal from the fob, the same way a wineglass begins to vibrate violently and is eventually shattered by a musical note it resonates with.

Another theory argues that the brain itself, that salty semi-solid mass, powers up the original signal by virtue of its strong conductivity. In this scenario, holding the fob to your head creates a much larger volume (living tissue) to act electrically in concert with the fob.

Dr. George Eleftheriades of the University of Toronto has pointed out that as soon as the fob is in contact with the head, they're linked, and the tiny antenna in the fob has suddenly become a much larger antenna. Because every antenna is slave to its own shapes and angles, the position of your head or the fob could be crucial. He suggested trying this: with the fob in your hand, stick your arm straight out in front of you, then rotate slowly to see if you hit a sweet spot.

It must be true, though, that the wateriness of the brain plays a role, because I was able to increase the distance my fob signal traveled by pressing it not against my head, but against a full hot water bottle balanced *on* my head. It boosted the range to 60 meters (about 197 feet)—double the normal. So booster cables, tire jack, hot water bottle—all should be in the trunk of your car!

What is the stopped clock illusion?

STAND IN FRONT OF A MIRROR and focus on the right eye of your reflection. Now switch and focus on your left eye. Then back and forth as many times as you like. What did you notice?

Not much, right? But it took time as your gaze shifted from one eye in your reflection to the other. Neuroscientists estimate that it takes tens of milliseconds, maybe up to a tenth of a second. During that brief time your eyes weren't really focused on anything; you should have seen a blur of movement before they settled on the new target. But you didn't. You don't *ever* see that. That missing blur is part of the stopped clock illusion.

TRY THIS AT HOME! Start the timer on your oven or your microwave or the stopwatch on your phone. Then shift your gaze (without moving your head), take some time to look around, then glance back at the timer. What do you notice? Most of the time, it will seem as if the number you first see pauses for a moment, then the count resumes. Every subsequent number ticks off regularly after that. No matter how many times you do this, you will still get the effect: the first number seems stuck. This is the stopped clock illusion, too.

So what's going on?

The phenomenon has been given the name *chronostasis*, the stoppage of time. And there are some pretty cool experiments that have shed light on it, suggesting that our brains are willing to play with our perception of time.

In 2001, Kielan Yarrow of the University of London and his colleagues published a set of experiments that suggested that the paused countdown is an illusion generated by the brain to cover up what would be a confusing blur of activity caused by the movements of the eyes.

They first created a setup in which participants stared at a point on the left side of the computer screen, then switched to a counter on the right. Their eye movement started the counter, which ticked away a second at a time: 1, 2, 3, 4. The trick was that the first number, 1, didn't necessarily last exactly one second but could be more or less—that was controlled by the experimenters. They could dial the first number up or down until subjects claimed it lasted exactly one second, just like the next three numbers. They found that when the number 1 lasted only 880 milliseconds, 120 milliseconds less than a full second, subjects thought it had lasted a full second. They thought that first number had lasted longer than it actually did. Had the experimenters dialed it up to a full second, it would have seemed to be significantly more, which is exactly what happens when you glance at the timer in the kitchen. (The illusion works best if the number your eyes land on has just changed—if it's already been on for as much as half a second, the illusion won't be as compelling.)

What became clear was that the farther the eyes had to move (and so the longer they took to arrive at the target), the longer the first number appeared to pause—in other words, "stalling" just long enough to paper over the blur of movement. Yarrow also ascertained that the illusion was linked to the eye movements themselves, because if the target was shifted into view but the eyes stayed still, the effect disappeared.

So when your eyes move from one target to another, there is no blur because your brain covers it up with an image of the target once your eyes have landed on it. The most startling thing about this is that there's a bit of time travel involved, because your brain is, in a sense, making you ignore the blur by focusing on something you've seen later—back-timing.

Of course this doesn't just happen in the cloisters of a psychology lab. Nor does it just happen when you glance at your microwave. It must be happening over and over during every waking hour in myriad situations. And there's evidence that it applies to more than just vision.

Imagine you're holding your cell phone to one ear, listening to the phone ring on the other end. Now switch to the other ear. If you do that a few times you'll notice that the pause before the next ring seems unusually long immediately after you switch ears. It's the same chronostasis effect, except this time the brain is masking a gap in hearing rather than a gap in vision. There are also experiments showing that when you reach for an object, you might backdate the time of the arrival of your hand, thinking you touched it before you actually did. These last two examples seem to suggest that your brain is making some temporary change in the speed of your internal biological clock, leading you to think less time has passed than actually has.

These experiments paint an unexpected picture of your everyday life: your personal time is constantly shifting. You think you're seeing everything that actually happens, but your brain is busy inventing it all for you.

If an asteroid was headed for Earth, could we shoot it down?

THE ANSWER IS . . . a definite *maybe*.

As you know, Earth is not the only thing circling our sun. So, too, are other planets and moons—as well as millions of asteroids and comets. Most of these asteroids and comets follow an orbit that runs between that of Jupiter and Mars, but occasionally some escape and hurtle toward Earth instead. So far, more than 17,000 of these wanderers have been tracked.

You act like the world revolves around you.

Scientists have good reason to try to keep tabs on all of this outer space rubble. As you may remember, the crash of one of these asteroids into Earth has been blamed for either killing off the dinosaurs, or at least helping them and many other creatures on their way to extinction. That asteroid was about 12 kilometers (about 7.5 miles) across and hit what is now the southern part of the Gulf of Mexico around 65 million years ago.

Another of these potentially deadly impacts happened in June 1908, when an object guessed to be about 40 meters (about 130 feet) across exploded in the air over the Tunguska River in Russia, flattening 2,000 square kilometers (772 square miles) of forest. Luckily, even though the blast was powerful enough to knock people off their feet, no one was killed. (That part of the country is very sparsely populated.)

Then in February 2013 a rock estimated to have been about 18 meters (almost 60 feet) across exploded above the Russian city of Chelyabinsk. Some scientists think the meteor may have briefly outshone the sun. Extraordinarily loud, the blast sent more than a thousand people to the hospital, most of them with cuts from breaking glass as windows were shattered by the shock.

As the examples above suggest, when it comes to flying space rocks, size matters. Obviously, if an incoming object is more than 100 meters (around 330 feet) across—roughly the size of a football field, in other words—we need to pay very close attention to it.

Science _Fact!_ *Our solar system began as a vast disk of dust and gas that then coalesced into the planets and their moons. Much of the original material never made it to the planetary stage, but collided over and over until it was reduced to small hunks of rocks and minerals—asteroids—or dust and ice—comets.*

Of the 17,000 asteroids tracked, more than a thousand have been labeled potentially hazardous—that is, big enough to destroy a city and with an orbit that will bring them within 7.5 million kilometers (more than 4.5 million miles) of Earth's orbit. Luckily for us, at this moment, none of these seem to be heading toward us.

But that might change. Then what? Intercepting an asteroid of any kind at any distance with any technology is going to be very challenging. What we would be able to do would depend on how close it was and how much time we had. Some of the techniques being studied are designed to ease the asteroid into a different orbit, causing it to miss us, while others are a little more violent—they aim to blow it up before it gets here.

Believe it or not, spray-painting has been suggested as one method of steering an asteroid aside. This involves sending a spacecraft to the asteroid (something that has already been done more than once) and spraying the surface with highly reflective paint or other material. This newly reflective surface would absorb and then emit photons (light from the sun) differently than the asteroid had when it was dark. The tiny—very tiny—pressure exerted by these photons could, over a long time, change the asteroid's orbit. But this would likely not be useful for large asteroids.

A similar idea, and the one that answers the question in the title of this chapter, is to send several spacecraft swarming around the asteroid. These would focus their lasers on the asteroid surface. The pressure of the laser light over extended periods of time should alter the asteroid's course. This project was originally called Mirror Bees because researchers had hoped to reflect sunlight to push the space object, but it turns out that lasers work much better.

Large asteroids need even more powerful techniques. The gravity tractor is a proposed spacecraft that would hover next to an asteroid, exerting a slight gravitational pull that would build over the years to nudge the asteroid into a new orbit. The gravity tractor would be a huge craft, requiring equally huge amounts of fuel to sustain its mission. And that's not the only challenge to this idea. The spacecraft would be firing its engines all the time to keep hovering over the asteroid. To do that most efficiently, the rocket thrust would be directed at the asteroid itself, pushing it away—exactly opposite to the goal of attracting it with the spacecraft's gravity.

A technique that would look similar—until arrival—would be an impactor spacecraft. This vessel wouldn't stop, but would crash at top speed into the asteroid. The impact would be designed to disturb the target's orbit enough to send it harmlessly by Earth.

And then there are the even more violent approaches. Shaving bits of an asteroid off with a high-powered laser has been suggested as a way we could defend ourselves, especially if there's not much time before a chunk of space rock hits. But the ultimate strategy would be to send a nuclear warhead to the asteroid and blow it to pieces. Russian scientists estimate that a mere 3-megaton bomb, trivial in today's nuclearized world, could blow an asteroid the size of the 2013 Russian one to bits. You can imagine there would be debates over the very use of a nuclear weapon, let alone its launch. There are also concerns that it would succeed—and blow the asteroid up into hundreds or thousands of pieces, some of which would follow the same orbit as the original asteroid and would still be big enough to do serious damage to Earth.

Did You Know . . . There is a spacecraft being designed right now called the hypervelocity asteroid intercept vehicle (HAIV). It's a way of delivering a nuclear explosion below the surface of the asteroid rather than in nearby space. A subsurface explosion would deliver much more destructive energy, but the nuclear weapon won't dig its own hole. Instead it is paired with a lead spacecraft that hits the asteroid at very high speed, digging a shallow crater into which the nuclear device lands and explodes. This technology is perhaps the only one that could intercept an asteroid with lead times of less than ten years.

All these techniques hinge on discovering the incoming space object soon enough. Estimates today are that it might take ten to twenty years to develop something like the gravity tractor, so tracking and identifying should be a high priority. The required sky survey is proceeding, but slowly.

How do we know Earth isn't flat?

IT DOES SEEM UNBELIEVABLE, but there are people in this world who still think Earth is flat, defying more than two thousand years of evidence that it is actually a sphere.

The latest believers to make headlines were the rapper B.o.B and "Mad" Mike Hughes in California. B.o.B started a GoFundMe campaign to use satellites to prove his point. The fundraising hasn't gone all that well. Mike launched himself in his own rocket to show that all this "Earth is a globe" stuff is wrong and, though he quickly came down, he is apparently planning a much higher altitude flight. But even if either one had been wildly successful, they'd have nothing but photos to prove their point, the same kind of photos that show the earth is round. Flat-earthers call those "fake," but they'd have to claim that Mike's and B.o.B's were the real thing.

Science _Fiction!_ *Some people who deny that the earth is spherical believe it is "round" but also flat—like a pizza. In fact, one of the main flat-earth arguments is that the world is round, with the North Pole at the center and a huge wall of ice (Antarctica) running around the edge.*

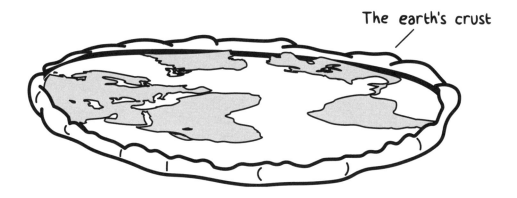

The earth's crust

So let's be gracious and say that doubters who believe the earth is flat are eccentric. The real question is this: are there any observations that would support the idea of a flat earth? I'm not talking here about casually looking across the prairies and saying to yourself, "It *looks* flat." I mean an actual experiment.

Most flat-earthers cite one such an experiment. It's called the Bedford Level, designed and executed by a man named Samuel Birley Rowbotham in the late 1830s. Rowbotham, who wrote under the pen name Parallax, suggested that if the earth were actually a sphere, then a six-mile (nearly ten kilometers) section of the Old Bedford River in England would provide the perfect setting for a test. The Old Bedford was really more of a canal than a river, and wasn't just straight, but also usually very calm. Rowbotham pointed out that over that 6-mile (9.6-kilometer) length, if the circumference of the earth is really 25,000 miles (40,000 kilometers), the curvature of the earth should be obvious. In his first experiment, Rowbotham held a telescope eight inches (20 centimeters) above the water and watched as a boat with a three-foot-high mast (close to a meter) sailed slowly away from him. He claimed that he was able to watch the boat all the way to the six-mile point, even though, if the earth were curved, the boat should have disappeared over the horizon.

In a second experiment, he placed five-foot-high flags at each mile mark along the canal and looked at them through a telescope also standing exactly five feet high. He claimed that every flag along the canal appeared to be at the same height, again proving a flat earth.

Rowbotham argued his case for twenty years or so without attracting much notice, until a man named John Hampden, also a flat-earth believer, promised to give 500 pounds sterling, a huge amount of money then, to anyone who could provide a convincing demonstration of what he called the "rotundity" of the earth.

The challenge was taken up by Alfred Russel Wallace. You may not be familiar with that name, but Wallace was an excellent scientist, so excellent that he came up with the theory of evolution at exactly the same time as Charles Darwin. He's known for that, but not for the unfortunate flat-earth wager. Unfortunate because he both won and lost.

Wallace's demonstration took place at the same six-mile stretch of the Bedford River. He installed three thirteen-foot (almost four meters) poles, one at each end of the canal and one in the middle. He argued that if the three poles were viewed through a telescope, the middle pole would be higher by about five feet (one and a half meters), because the three poles would trace the curvature of the earth. He was right, won the bet, but then was harassed and even threatened by Hampden for the next several years. In the end his life had been made more miserable by winning the bet.

But how had Rowbotham got it so wrong in the first place? The standard explanation is that he failed to take into account that light bends as it travels through the air. If there's a layer of warm air above cooler, then light passing through that air will bend toward the ground instead of continuing straight. This means that what you can't see over the curve of the earth will be brought into sight, as if the horizon is curling up all around you like cheap wallpaper. Some have suggested that when the Vikings reported seeing giant whirlpools in the North Atlantic, this is what they were seeing.

Rowbotham could indeed have seen the flags standing at the same height if, on that day, the curvature of light rays was roughly the same as the curvature of the earth: they'd neutralize each other. Or at least come close.

So the Bedford Level, supposed to be the best demonstration of a flat earth, is miserably flawed. Yet here we are, nearly a century and a half later, and there are flat-earthers of many different varieties still around. They may believe the earth is flat, but there's just no evidence.

How can some people wake up just before the alarm goes off?

ARE YOU ONE OF THESE PEOPLE? You have to get up at 6:30 a.m., so of course you set the alarm, but then you wake up, apparently out of a sound sleep, at 6:28 a.m. It seems astounding—but is it just some sort of coincidence? And if not, if you are able to set some sort of internal clock, what exactly goes on in the body to make this possible?

Surely some on-time waking *is* a coincidence. Anyone who is waking up regularly throughout the night might well have one of those wakings land very close to the morning alarm time.

But there are more than a century's worth of studies that suggest that some instances of waking up on time without an alarm are more than just a coincidence. My favorite example

I have a fine-tuned internal clock.

of a dedicated investigator of this phenomenon is a doctor named Winslow Hall, who in the 1920s was apparently able to wake up within fifteen minutes of the desired time in more than 50 percent of the one hundred plus trials he ran. And he woke at the exact minute eighteen times.

Of course something a little less personal and a little more scientific will be necessary to satisfy the skeptics out there. Such studies have been done, and they're all pretty consistent, arriving at the conclusion that yes, some people can wake up more or less at the prescribed time, and some can't. For those who have the ability, motivation is important. We all know this intuitively: it's much easier to wake up when you have an important event, like getting to the airport on time or taking an exam. It's also more likely in these situations, however, that you will wake up several times, and as I mentioned earlier, when multiple awakenings occur, it's hard to know if one that happens just before your alarm goes off is just a lucky chance or not. The truly useful ability—to wake up once, right on the dot of the alarm time—appears to be pretty rare.

For those who have this gift, what is the cue to wake up? Hall reported a dream where he saw a hand-lettered sign, black letters on a pink background, that read TRUTH, which woke him up. In other experiments in the 1920s, some participants reported they had heard voices saying things like "It's time, you must wake up now." That would be helpful! But for most, it appears as if they just woke up, no message necessary.

The fact that some have either heard or seen cues (in their heads) to wake up points to one consideration: the stage of sleep. Generally speaking we spend the first part of the night in the deepest phases of sleep, but as the night wears on, we spend more and more time in REM (rapid eye movement) sleep, a lighter sleep when almost all our dreams occur. Waking out of REM sleep is easier than waking out of deep sleep. So it makes sense that we often wake while dreaming and more often toward morning. But at least so far there's not much evidence that dreams, or their content, directly trigger waking up.

The fascinating thing about waking up on demand is that it requires two things: the brain's ability to store the message before going to sleep and trigger it later at the right time, *and* a biological clock on which that brainwork depends.

It's no surprise that a biological clock is involved. Our bodies run on several of them. These operate 24/7, controlling a full sequence of chemical events throughout our systems. (Jet lag is the result of disturbing those timekeepers.) Happily we now have some idea of how the biological clock makes self-waking possible. In 1999 a team at the University of Lübeck in Germany challenged volunteers in a sleep lab by telling them they would be wakened at either 6:00 a.m. or 9:00 a.m. the next morning. While the subjects were asleep, blood samples were taken every fifteen minutes. They revealed that in the group who had been told they were to be wakened at 6:00 a.m., the stress hormone adrenocorticotropin began to rise at about 5 a.m. The group who thought they were going to be able to sleep until 9:00 experienced no such preawakening hormonal surge.

Hormones might explain coincidental waking as well. If an important and stressful event necessitates waking early, stress hormones might be at higher levels through the night, causing multiple awakenings, the last of which could convince people that they had miraculously woken up at just the right time.

That a chemical basis for wakening has been identified puts this whole puzzle on firmer ground, but there's still an unanswered question. How exactly does the brain count the hours through the night until waking time, including beginning hormone release an hour before, while you're unconscious? Well, we know that something said immediately before falling asleep can be remembered in the morning; so could a wish to wake up on time. It's also true that our judgment of the passage of time is about as accurate when we're asleep as awake. Put the memory and timekeeping together and you have a system that can do this. It does still seem uncanny, though.

Do bacteria live on money?

It seems like every week we learn something new and significant about the human microbiome, the trillions of bacteria that live both on us and in us.

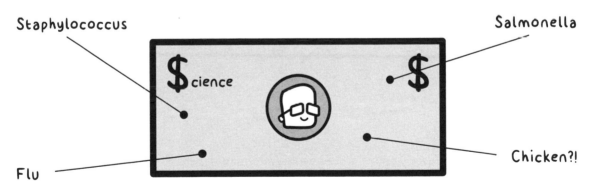

There are more of these bacteria than there are body cells (that is, we're more bacterial than we are human), and the vast majority are helpful. But some are infectious. Those potentially dangerous bacteria are mostly held in check. Bacteria that encounter us from outside our microbiome, however, can be a different matter. We aren't set up for them, and they may get out of control and create an infection.

Infectious bacteria have been the architects of fearful epidemics. The Black Plague (or Black Death) in Europe from 1347 to 1351 killed more than 100 million people, almost half the population of Europe. Many bacterial species can now be controlled by antibiotics, but some bacteria are stubbornly resistant. This is why we're so interested in (or concerned about) the bacteria around us.

And they're everywhere: on doorknobs, keyboards, steering wheels. In a typical day you'll be in continuous contact with bacteria. Let's say you want to take some money out of an ATM—a routine activity. Are there bacteria on the keyboard? Or, even more disturbing, on the money you withdraw from the ATM?

The concern about "dirty money" is long-standing. In fact, during the late nineteenth century one of the first studies was done, using Hungarian banknotes from 1848, and indeed there were bacteria on them.

In a 1972 paper, Drs. Berel Abrams and Norton Waterman described looking for bacteria on $13.47 worth of American coins and a few dozen $1 and $5 bills.

Abrams and Waterman were funny guys: among their recommendations for protecting yourself against infection were to keep your money in the bank, keep your hands out of your pockets, get rid of your money quickly, and campaign against piggy banks! They also offered to accept money from you if you feared it might be contaminated, and promised to dispose of it safely. Between jokes, Abrams and Waterman managed to identify six types of bacteria on the money. They did this by swabbing the coins or a corner of the bills, putting that swab into a nutrient broth, and letting the bacteria grow until they could be identified.

In a follow-up study in 2002, scientists tried to grow bacteria from a set of 68 dollar bills they had collected. They found bacteria on 64 of them, including lots of human-skin-related bacteria. While bacteria that would cause disease in healthy humans were relatively rare (on only five of the bills), fifty-nine bills had bacteria on them that could pose a threat to people in hospitals or those with compromised immune systems.

All these studies were limited by technology. The problem is that many bacterial species simply refuse to grow in nutrient broth or on a petri dish in the lab. Using only those methods guaranteed that the numbers of different kinds of bacteria were dramatically underestimated. (Considering that millions of bacterial cells are likely to be transferred at every touch, this is a significant issue.) A better way is to collect samples of the DNA that living or dead bacteria have left behind, and then identify the species responsible. Prospecting for DNA like this instead of trying to grow living cells has vastly expanded our knowledge of the bacterial world.

For instance, scientists from New York University harvested DNA from the keys of ATMs in three New York City boroughs: Queens, Manhattan, and Brooklyn. Then it was simply a case of comparing that DNA with known bacterial genomes (these are on file) to find out which species left their genes behind on the ATM keys.

This New York University study reported a large variety of bacteria, but also uncovered a variety of single-celled animals, like amoebas, and traces of human food as well.

In all these studies there were some disease-causing bacteria found on the keyboards, but usually nothing to worry about. Many of the bacteria on the keys were types commonly found on household surfaces in kitchens or bathrooms or on pillows and TVs, each of which has its own typical bacterial population. It's likely people transfer as many bacteria to the ATM as they pick up from it. Although, just to maintain an air of mystery, the vast majority of bacteria found in this study were from an "unknown" source.

More interesting was the fact that ATMs in different parts of New York City harbored different mixes of DNA from different species. In Chinatown, there were traces of fish and mollusks, but in Harlem, evidence of chicken. There are more fish restaurants in Chinatown, so probably people who had just eaten fish then stopped at the ATM. Same for chicken in Harlem. (Previous studies have shown that DNA from food is spread all over New York by the subway.)

What about the money? The most recent DNA analysis was published in 2017 and identified an incredible 549 species on paper money. It demonstrated that many of these were indeed alive (and not just broken scraps of DNA left behind) and that there were significant differences between what was found in winter and in summer. The study also demonstrated why caution should be exercised when identifying DNA: the discovery of "wolf" and "wild boar" DNA on paper money in New York City was more likely a sign that the handlers of the money had come in contact with pigs or dogs, domestic versions that share some of their DNA sequences with their wild cousins, rather than the possibility there are feral packs roaming the streets of Manhattan.

One final note that might be cheering or not depending on your attitude. Paper money in the United States (where the studies quoted here were carried out) is a mix of cotton and linen. "Paper" money in Australia and Canada isn't paper at all, but plastic. Bacteria survive much, much longer on the cotton-linen mix, so if you're apprehensive about handling bills rich in microbial diversity, you'd be happier in Australia and Canada. If, however, you'd don't mind stuffing your pockets with filthy lucre, the United States is the place to be!

History Mystery

What were the Hanging Gardens of Babylon?

THE HANGING GARDENS OF BABYLON were one of the traditional Seven Wonders of the Ancient World. They were, however, a mysterious inclusion. For one thing, the Hanging Gardens were a geographic outlier: the other six could easily be visited if you took a cruise along the eastern coast of the Mediterranean. But to get to the Hanging Gardens you'd have to head overland to what is now Iraq, more than 2,000 kilometers (1,300 miles) away. But the isolation of the gardens wasn't the only strange thing about them. Unlike the other Seven Wonders, they weren't astounding feats of architecture or engineering. What's more, they weren't hanging. And they might not even have been in Babylon!

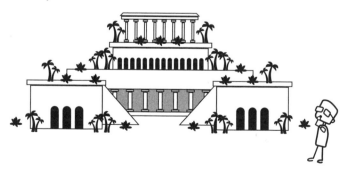

Did You Know . . . The complete list of the Seven Wonders of the Ancient World is the Great Pyramid at Giza (the only one still standing), the Lighthouse at Pharos, the Colossus of Rhodes, the Tomb of King Mausolus at Halicarnassus, the Temple of Artemis at Ephesus, the Statue of Zeus at Olympia, and the Hanging Gardens of . . . uh . . . Babylon. For now.

The Hanging Gardens were thought to have existed in the seventh and eighth centuries BCE and been built by the great Babylonian king Nebuchadnezzar. But for centuries, archaeologists have searched for the ruins of the gardens without success.

Yet it seems as if the Hanging Gardens of Babylon shouldn't be hard to find. The "where" is the ancient city of Babylon, in southern Iraq, not too far north of the Persian Gulf. Its ruins are well known to archaeologists.

Nor should the "what to look for" pose any serious problems either. While the descriptions we have are often fragmentary and vary from one to the next, there are consistencies. The Hanging Gardens were said to have been built next to the royal palace, on a series of stone terraces (not bricks), to imitate a mountainous landscape. They displayed trees of every description, and "enough soil was heaped to allow the biggest trees to take root." Somehow, miraculously, water for the trees and flowers was transported to the highest levels of the terraces.

The descriptions were written two or three hundred years after the gardens had supposedly fallen into ruin and could contain much exaggeration, but nothing that even hints at the magnificence and grandeur described in the ancient writings has ever been found in the archaeological digs of Babylon. Even if, as some have suggested, the gardens were really only a relatively minor rooftop affair, they likely would have baked in Babylon's summer heat. And while the gardens were said to stand next to the river to make watering more convenient, the only likely site near the river would have shaded the plants for most of the day.

Dr. Stephanie Dalley of Oxford University, a scholar intimately familiar with the writings of the ancient Near East, wondered if the difficulty locating the Hanging Gardens might lie in the fact that everyone was looking in the wrong place. She knew that the writers of the time often confused the king of Babylon, Nebuchadnezzar, with King Sennacherib of Nineveh, another great city about 500 kilometers (300 miles) to the north. Those writers also confused the two cities and even the rivers on which they were built: Nineveh on the Tigris, Babylon on the Euphrates.

The writings are admittedly unreliable, but that alone wouldn't be enough to argue that we should be calling them the Hanging Gardens of Nineveh. But Dalley had more.

Astronomical texts from the time specify a length for the longest day of the year at Babylon, but while the numbers are precise, 14 hours and 24 minutes, they're not correct for the latitude of Babylon, but for a location much farther north.

The astronomers of the time were too skilled to make an error of that magnitude, and while the northerly location is still not a perfect fit for the latitude of Nineveh, this is more evidence that the name Babylon didn't always refer to the city of that name.

Science Fiction! Name confusion is one explanation for the missing Hanging Gardens, but there are other, more bizarre theories. Author Immanuel Velikovsky became famous in the 1950s for suggesting that catastrophic events mentioned in the Bible and other ancient texts were real, caused by an object from space sideswiping the earth and creating chaos. The Babylonian latitude discrepancy, he argued, was proof that after a close brush with the planet Venus as it hurtled through the solar system, Earth had tilted by exactly that much, moving Babylon farther south. Imaginative, but not convincing.

Sennacherib, king of Nineveh, described his own magnificent gardens (most kings of the time bolstered their image with the newest and greatest), but more important, mentioned a new invention that was used to water them. Traditionally water for irrigation was scooped out of a lake or river with something called a shadoof, a long pole balanced horizontally, with a basket at one end and a weight at the other. The basket was dipped in the water, the pole swung around, and the water spilled where it was needed. But Sennacherib described a process for casting cylinders and spirals out of bronze or copper for lifting water.

There is indeed a device for lifting water that is a combination of a cylinder and a spiral: the Archimedean screw. A spiral structure is fitted tightly into a cylinder; when one end is dipped into the water and the spiral rotated, the water is lifted to the top of the cylinder.

Archimedean Screw

The only hitch is that Archimedes is credited with the invention of the Archimedean screw—no surprise there—but he did it four centuries later!

It's hard enough sometimes to figure out who the real inventor of a modern technology was, let alone trying to do that for an invention from 2,500 years ago.

Stephanie Dalley's suggestion that the Hanging Gardens were actually at Nineveh is intriguing, but not all archaeologists agree. And it is likely going to be difficult to find additional evidence for or against the idea: the city was pretty much destroyed 2,600 years ago.

Science Fiction! One persistent idea about the Hanging Gardens is that they were built on a ziggurat, a brick tower with a tiered structure—like a wedding cake. Some ziggurats were enormous, up to seven levels, and accessible only by either a series of ramps on one side or a spiral ramp circling the structure. Sir Leonard Woolley, a legendary British archaeologist and the discoverer of the ancient city of Ur, claimed that a ziggurat there had holes punched in the brickwork to drain water from plants. It turned out, however, the holes had a different purpose. No ziggurat has ever been found that might have held plants in any profusion at all, but the idea of gardens "hanging" down or suspended from ramps on a ziggurat (like spider plants in a 1970s office) has been cemented in the public mind.

Part 4
Mystery Meals

Can some people taste more (or better) than the rest of us?

THE DISCOVERY OF SUPERTASTERS is only one of the many surprises in the science of taste over the last few decades.

For one thing, until recently it was thought we could detect four different tastes—sweet, sour, salty, and bitter—and each had its own location on the surface of the tongue. But it turns out that both the number of tastes and the idea that they're segregated on the tongue are wrong.

First, there is a fifth taste, savory, or umami (the Japanese word for "pleasant savory taste"). Soy sauce is rich in umami. Second, the idea that we taste sweet at the front of the tongue, salt on the sides, sour behind salt, and bitter at the back is

Mmmmm

misguided (despite being unquestioned for a century). All five tastes can be detected anywhere on the tongue that there are taste buds.

We now know that the tongue, the roof of the mouth, and the back of the throat are studded with structures called papillae. They're three different shapes: tiny towers, mushrooms, and slits in the tongue. The taste buds are embedded in them, and individual taste receptors, up to 100 per bud, are tucked inside. When stimulated by anything sweet, salty, bitter, sour, or savory, a receptor fires off a signal to the brain, alerting you to the presence of something with that taste. If a substance contains all five, you taste all five.

Taste is, however, more than papillae and buds: it also includes odor, temperature, and texture of foods. But taste buds play a crucial role. They are often described as sentinels, as their job is to alert us to the value—or danger—of things that we put in our mouths. Our hunter-gatherer ancestors constantly encountered new and unfamiliar food plants and animals that had to be evaluated. Animals that eat just one thing—for example, pandas (bamboo) or koalas (eucalyptus)—have a narrow range of tastes. Cats, strict carnivores, are unable to detect sweetness. But because we are omnivores, we need an elaborate range of taste detectors, especially for potentially hazardous substances.

Mmmmmm

So our tasting abilities help us choose foods that we need to survive. Sweet receptors signal the presence of valuable sugars, while savory tastes are associated with meats, especially those which have been cooked or fermented. As omnivores we're attracted to salt, rare in fruit and vegetables. By contrast, toxic chemicals are often bitter, and we shun those tastes. (The challenge for hunter-gatherers in the past was to get enough sugary and salty foods; the challenge in developed countries today is to resist them.)

So we all have a pretty impressive tasting apparatus, but what is a supertaster? The precise definition is someone who is extremely sensitive to the bitter chemical propylthiouracil, but that is only a tiny part of their story. Supertasters, about 25 percent of the population, experience a wide variety of tastes much more intensely than the rest of us. Bitters are bitterer; sweets are sweeter. And that's because of the sheer number of the mushroom-shaped papillae on a supertaster's tongue. In fact, supertasters have about twice as many of these as your "average" taster.

 TRY THIS AT HOME! Are you a supertaster? Make a hole about the size of a notebook hole-punch through a piece of plastic wrap or waxed paper. Next, swab your tongue with a bit of blue food coloring. Then place the wrap over your tongue so the hole covers a space near the front. Count the number of large taste buds you see. If you can count 30 or more taste buds within the hole, you may be a supertaster.

Why are some people so sensitive in the taste department? Supertasting might have evolved as a protection against toxic substances. And the fact that we're not all supertasters might be because it's an advantage only in environments particularly rich in toxic foods, but is superfluous in others.

Finally, if you're longing to be a supertaster, keep this in mind: there are some bitter foods, like coffee or kale, that people find delicious. So you might end up missing out. And while picky eaters are not necessarily supertasters, many supertasters are picky eaters. And that just means more broccoli for the rest of us!

Did You Know . . . While bitter taste buds help protect us from dangerous toxic substances, they can't keep you totally safe. One of the world's deadliest mushrooms, the death cap, *Amanita phalloides*, apparently tastes delicious, but kills you within a week. The Roman emperor Claudius may have died from amanita poisoning in 54 CE.

Most animals can eat lots of raw meat, so why can't we?

THE ANSWER TO THIS QUESTION gets at some serious developments in human evolution, but the question needs to be tweaked a bit. For one thing, we do eat raw meat—steak tartare and sushi being the most common examples. But obviously they're not nearly as common, at least in North America, as cooked meat. The other consideration is that we mostly cook our vegetables, too, and while that isn't the focus of the original question, it *is* part of the answer.

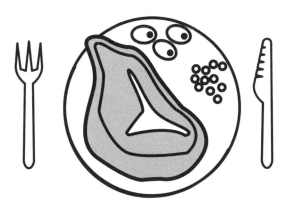

The risk of eating raw meat or fish is the chance that it is contaminated by bacteria or other microorganisms that could make us sick. Whether it harbors them or not depends on the cleanliness of preparation, the amount of time exposed to the air, and even, in the case of ground beef, the number of cows that contributed (unwillingly).

Animals, birds, and other species can tolerate meat contaminated with bacteria and microorganisms because they have an acid-rich, antimicrobial stomach and gut. Even then, though, they can be infected by parasites lurking in the flesh they consume. But for humans, the only safe way to consume tainted meat is by cooking it and thus killing the organisms. That is why we are urged to cook meat until the internal temperature reaches 165° F. or 74° Celsius, accurately measured by a meat thermometer.

Did You Know . . . That a package of ground beef at the supermarket might contain meat from as many as 1,000 cows!

If all the necessary precautions have been taken, however, it is safe to eat uncooked meat. But there might be reasons other than the prevention of illness that our ancestors began to cook meat. First, of course, they had to harness fire, and there is some evidence this happened as much as a million years ago. Once campfires existed, why not toss the meat in? Possibly they continued to do so after their initial experimentation, not because it tasted better—people who eat raw meat today think it's more flavorful—but because it made the meat easier to chew and digest. But in doing so, they made a much more important step: they grew their brains!

In the short term, cooking, because it breaks down difficult-to-digest components of food, would have meant less time devoted to eating, freeing up time for other activities. But the long-term effects were likely much more important.

The architecture of our heads would have changed, because our jaw muscles, teeth, and gut, with less work to do to get through dinner, could shrink. But cooking didn't just affect the way we looked. One of the most astounding things about our brains is that they consume 20 percent of the energy we take in even though they represent only about 2.5 percent of our weight. By comparison, the great apes operate on half that amount.

Brazilian scientists Suzana Herculano-Houzel and Karina Fonseca-Azevedo calculated that before we started to cook, it simply wouldn't have been possible, or at least practical, for early humans to spend the eight to nine hours every day necessary to gather and take in enough calories to grow and maintain our big brains. They argue that the fact that chimps, gorillas, and orangutans don't cook limits their energy intake and therefore the size of their brains.

So cooking meat could have helped make us smart (although it doesn't follow that barbecuing today will make you even smarter). But it seems that the advantages of cooking are not limited to meat. Harvard paleoanthropologist Richard Wrangham has suggested that vegetables, roots, and tubers would also have been much more digestible if cooked first, and even though they are not as energy-intensive as meat, spending less time consuming them would also have given our ancestors an advantage over other primates living at the time.

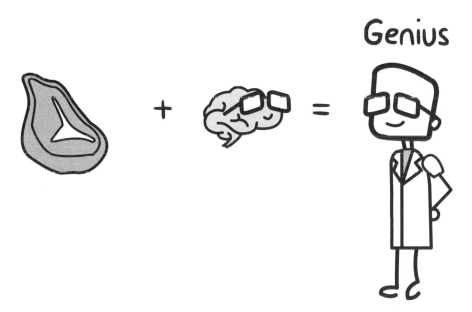

Neat theories, but there are sticking points. For cooking to have accelerated brain evolution, fire would have had to be tamed nearly two million years ago. But the most reliable dates for the first cooking fires are much more recent, long after the growth of our big brain had begun.

Did You Know . . . Vultures are the champions of eating dangerously. They scavenge rotting carcasses, which of course contain bacteria and their toxic by-products. But their guts take care of all that. More than 500 species of bacteria can be found on a vulture's face, but in the bird's stomach, there are only about 70. And bacteria that are extremely dangerous to us live happily in their gut.

Also there's evidence that slicing and pounding meat would have substantially cut down the effort required to eat it even before cooking began.

So are we smart because we started barbecuing, or do we barbecue because we're smart?

Why do cola and mints create a fizzy geyser?

You might have seen the "Coke and Mentos" demonstration: open a bottle of Coke, throw some Mentos in, and stand back! An enormous geyser erupts from the bottle, sometimes gushing as much as several meters in the air.

The answer to the question of what the heck is going on with this is more subtle than you might think. It's not unlike shaking a bottle of pop, then opening it, but with much more chemistry.

For the moment, let's ignore some of the ingredients in cola, like sugar (or artificial sweeteners), flavorings, and colorings, and concentrate on the main players—carbon dioxide gas and water. As the drink is being bottled, carbon dioxide gas is introduced into the liquid. Under pressure and at low temperatures, the gas dissolves in the liquid to the point where the water has accommodated all the gas it can. In fact, once the bottle is capped, the gas dissolved in the drink and the gas in the air space above the drink are in equilibrium: there might be a small amount of gas traffic in and out of the liquid, but the total concentrations in both remain constant.

It's the peculiar nature of carbon dioxide in this situation that as long as the pressure is maintained (as long as the cap is sealed), nothing will change. However, as soon as the cap is removed, the pressure in the air space drops immediately and the carbon dioxide in that space escapes from the bottle. Now equilibrium has been lost, and the dissolved gas in the liquid starts to form bubbles that rise through the liquid and burst when they reach the surface. It's not a super-fast process, but it's continuous. As everyone knows, the longer you wait, the "flatter" your drink—in other words, the less carbon dioxide remains in the liquid.

That process of turning dissolved molecules of carbon dioxide into bubbles of the gas isn't simple. For one thing, when they're dissolved, they're surrounded on all sides by water molecules. But bubbles of carbon dioxide are collections of vast numbers of molecules of the gas, so somehow there has to be a way for them to collect. That usually requires a place for the carbon dioxide molecules to assemble, a so-called nucleation site. This can be a rough patch on the inside of the bottle or tiny impurities in the drink—any place that will allow the gas molecules to shoulder the water

I'm having an out-of-bottle experience.

molecules aside and cling together. You can see this much more clearly if you pour the drink into a glass: streams of bubbles rise from specific places on the inside of the glass. Those are nucleation sites. Once a bubble forms, it rises to the surface because it's less dense than the surrounding drink; on the way, it becomes its own nucleating site and so will collect even more carbon dioxide and grow.

That's the routine set of events that follow twisting the cap off a bottle of pop. Adding a Mentos (or several) changes the picture dramatically. The beauty of Mentos—at least in this context—is that the surfaces of the candy, as revealed by the scanning electron microscope, are landscapes of pits, rough edges, and jagged promontories—in other words, nothing like the apparently smooth candy you hold in your fingers. Such surfaces offer a wealth of nucleation sites, so when the candy falls to the bottom of the bottle, the number of opportunities for bubbles to form is multiplied by thousands, if not millions. If carbon dioxide molecules could celebrate, they would.

The result is obvious: the newly formed bubbles rushing to the surface overwhelm the ability of the bottle to contain them, and they explode into the air, carrying huge quantities of liquid with them. Chemists have investigated this in some detail and have identified some additional features that provide the most dramatic explosive effect.

One is that bubble formation might be a two-step process. The first wave is a set of bubbles forming on the surfaces of the Mentos, but the second is an additional surge of bubbles created by the turbulence set in motion by the first wave.

Additional chemical compounds usually found in soft drinks may contribute in another way—by preventing bubbles from growing too much. A large bubble offers less surface area for the addition of more carbon dioxide than does an equal volume of smaller bubbles, so forming as many as possible, but limiting the maximum size, would contribute to a bigger explosion. Warming the Coke up helps, too, because a warmer liquid is less able to hold carbon dioxide in solution.

Diet Coke works better than Coke containing sugar. This is due to the fact that aspartame, the artificial sweetener that replaces sugar, is a surfactant, something that lowers the surface tension of liquid. Surface tension holds water molecules together.

The gummy substance in Mentos does the same. The two together, by reducing the surface tension in the Diet Coke, make it easier for carbon dioxide molecules to break through the film of water surrounding them and form bubbles.

 TRY THIS AT HOME! If you want to *see* surface tension, the attraction of water molecules to each other, gently place a needle or pin on the surface of water in a glass. Surprisingly, the needle floats, because it is supported by the filmlike surface tension of the water molecules on top of the water. Now add a single drop of detergent; the needle sinks immediately! Detergent is what's called a surfactant; surfactants reduce surface tension, and in this case diminish the ability of the water to support the needle.

Fruit- or mint-flavored Mentos generate more powerful geysers than other Mentos or candies, probably because they have rougher surfaces, allowing for faster bubble formation, although in at least one experiment fruit-flavored Mentos worked faster and created an eruption that traveled much further. Because plentiful surfaces are so important in this reaction, you might expect that substituting Mentos with salt or sand would give about the same results, but neither is as effective as the candy. One possible reason is that Mentos descend faster through the Coke and so expose themselves more rapidly to carbon dioxide throughout the bottle.

For reasons that aren't clear to me, most of the research on bubbles in drinks to date has focused on champagne and beer, especially Guinness (why?). Thankfully, the chemistry of Coke and Mentos has redressed that balance a little.

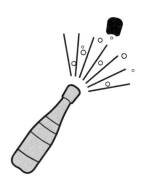

How does popcorn pop?

THERE ARE TWO WAYS of answering this question: the quick and dirty way and the deep dive.

The quick version is pretty straightforward. You begin with a special kind of corn. The creamy corn in a can will never pop, but a particular variety, *Zea mays*, will. Why? Because it has a thick outer shell on the kernel, called a pericarp. The inside, however, isn't much different from other kinds of corn. It consists of the seed, starch to help nourish the seed if the corn is planted, and water. When a kernel like this is heated, the importance of the thickness of the pericarp comes into play. The water begins to boil, turns to steam, and builds up pressure inside the kernel.

At a critical point, right around 180° Celsius (325° Fahrenheit), the pressure becomes too much for the pericarp to withstand, and it cracks open. The starch, which has become spongy because of the heat, bursts out. Once released from the kernel's shell, it cools quickly

Quick and dirty.

Deep dive.

(partly because the pent-up energy has been used to expand it) and is ready to eat. Some kernels fail to pop, either because the pericarp is unusually thick or because somehow during the heating, the steam leaked out through the tip of the kernel and the necessary internal pressure was not reached.

Did You Know . . . North Americans consume 20 billion quarts of popcorn every year.

That's how popcorn works, but a closer look reveals details you might never have suspected. French scientists Emmanuel Virot and Alexandre Ponomarenko revealed those secrets in a study published by the Royal Society.

Three Stages of Popcorn

Kernel   Heated   Popped

After confirming that 180° Celsius is the crucial "popping" temperature, the two scientists addressed two other curiosities of popping corn: why it jumps and what causes the "pop" sound.

Their analysis showed that, on average, a piece of popcorn jumps anywhere from a few millimeters up to several centimeters (an inch or two), although I've seen much bigger jumps of a foot (25 centimeters) or more. A high-speed camera recorded that the force behind the jump is not, as you might have expected, caused by a sudden release of steam that turns the kernel into a rocket. Instead, a kernel resting on the surface of a hot pan becomes airborne when the sudden extrusion of the white, fleshy starch (the scientists call it a "leg") pushes the kernel off the bottom of the pan.

Of course some kernels won't launch at all if the leg explodes upward or sideways, but nonetheless, the power of the push, when it happens, is impressive. Considering it sends the kernel upward and sideways *and* sets it spinning, its performance is comparable to both that wizard of jumping, the flea, and the somersaulting ability of a human gymnast. The flea performs better, but let's be fair: it has taken millions of years of evolution to perfect its jump. It's in comparison to the gymnast that the popcorn kernel's marvelous abilities should be appreciated.

Science _Fact!_ *Traces of popcorn can be found in archaeological sites dating back several thousand years.*

To make a powerful somersault, the gymnast tucks in midflight, thereby reducing something called angular momentum (the same thing figure skaters do to accelerate their spins). The popcorn, however, encounters the disadvantage of actually swelling, not shrinking. Nonetheless, it generates roughly the same amount of rotation as the human.

Virot and Ponomarenko also examined the sound of popping corn by placing a combination of video camera and microphone next to a hot plate. They saw that the kernel cracked open first, the spongy starch of the kernel began to bulge out, but then, even before it hit the surface to launch the kernel, the microphone recorded the pop. So the sound isn't caused by the initial split of the hard outer shell, nor by the impact of the expanding starch hitting the surface, leaving only the release of steam as the cause of the sound. They compared this to a more familiar sound: that of a champagne cork being popped (they are French, after all). The sudden release of pressure inside the kernel (as in the bottle) does the trick. The "pop" is actually a few rapid pops, a few thousandths of a second apart, likely the result of different pockets of steam being released one after the other.

Who would have suspected that common popcorn could conjure up images of fleas, gymnasts, and champagne? Truly a wonder food!

This is so corny.

What makes
peppers spicy?

THERE'S A LOT TO CONSIDER IN THIS QUESTION. Besides the chemical answer to the "what" of spiciness, there are at least two other considerations: Why do humans like spicy food? Why are hot peppers hot in the first place?

First the "what." Capsaicin is the chemical that makes peppers spicy. The more capsaicin in the pepper, the hotter it is. Heat in peppers is measured by something called Scoville heat units, named after Wilbur Scoville. The original test involved human tasters who would sample a series of sugar water dilutions of the pepper until there was no "heat" detectable, but now the Scoville units are determined by more accurate chemical tests.

I'm a hot and spicy pepper, baby.

Did You Know . . . The spiciest five peppers are the Carolina Reaper, the Trinidad Moruga Scorpion, the ghost pepper, the Red Savina habanero, and the Scotch bonnet, tied with the habanero chile. The Carolina Reaper tops out at 2,200,000 Scoville units, the Scotch bonnet a mere 350,000. Compare those champion numbers to the tabasco pepper at 50,000 and paprika at 900. Bell peppers? Zero.

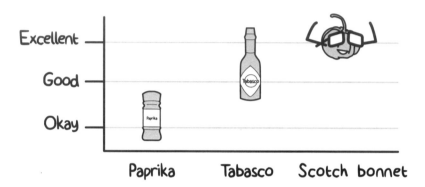

Scoville Heat Units

Excellent — Good — Okay —

Paprika   Tabasco   Scotch bonnet

There are countless YouTube videos of brave (?) individuals eating the hottest peppers known, and after watching a few of those, you won't be surprised to hear that the heat of peppers isn't really heat—it's pain. Capsaicin triggers pain receptors in the lips, tongue, stomach, and territories beyond. Somewhat strangely, that pain is perceived—if it isn't too extreme—as enhancing the taste of the foods containing the peppers. Of course seasoning food with tiny particles of that pepper is not the same experience as eating an entire pepper. Leave *that* to the YouTubers.

Science _Fact!_ *A surprise spin-off from the pain-inducing qualities of capsaicin is that applied carefully, capsaicin can be a painkiller. Apparently it causes defunctionalization—it overwhelms the pain receptors in that area and, paradoxically, offers extended pain relief.*

But if a hot pepper's essential impact on us is painful, why have we developed a taste for it? Some global surveys have strongly suggested that the use of peppers and an array of other spices is probably left over from a time when refrigeration was virtually nonexistent and the spoilage of food was much more common than it is today. Many spices, even common ones, have significant antibacterial activity. Garlic, onion, and allspice are the top-ranked; hot peppers rank twelfth.

Is it really true that there's a connection between the prevention of bacterial contamination and the use of spices? If the use of spices is mapped across the world, it's clear that countries with hot climates make routine use of spices (at least in meat-based dishes) much more often than countries in cold climates do. For instance, Ethiopian, Indonesian, and Indian cooking routinely make use of between seven and nine spices per recipe, in each case including hot peppers. By contrast, Denmark, Poland, and Norway, where for months you could leave food outside without worrying about bacterial contamination, average fewer than two spices per recipe, with not a hot pepper in sight. The hotter the country, the hotter the most common spices in that country's recipes are. (Even within countries, and China is a good example, regional spice use varies with differences in average temperatures.) Interesting that, just as onion and garlic are two of the most potent antibacterial spices, those two plus black pepper are the three most commonly used around the world. Hot peppers rank fourth.

Science Fiction! *One theory suggests that we use spices in cooking to cover up the odor of spoiled or decaying food. They might do that, but people can easily get sick from eating spoiled food, so anything that encourages them to do so isn't a very sensible strategy. It's also been suggested that spices cause people in hot countries to sweat and therefore help them cool down. Although hot peppers actually do that, very few other spices do, so that's not much of an explanation. Even the suggestion that spices contain beneficial trace nutrients doesn't explain the global distribution. And some such claims, like the one that turmeric prevents Alzheimer's disease, have no reliable data to support them.*

But why do some plants produce fruit (that's what the pepper is) that is so painful to eat in the first place? Well, capsaicin provides a defense for the plant, too. Any animal that likes to eat fruit will eat a hot pepper only once (unlike us humans). This may seem like a self-defeating strategy at first. The whole point of producing a fruit is to encourage it to be eaten so that after it has been digested, the seeds inside are pooped out on the ground, where they can germinate. If an animal will eat only one pepper, how can the plant reproduce successfully?

Ah, but there are birds, too. And birds are not sensitive to the irritation and pain of hot peppers. They could go Scoville crazy and not feel a thing. So the heat of the peppers might be a way of encouraging birds to feed on them while discouraging mammals.

Why is it to the pepper plant's advantage to exclude mice and rats? Apparently these rodents actually destroy the seeds when they consume them, whereas the birds digest the flesh of the pepper and poop out the seeds whole. And birds deposit them in places where they will germinate best: in the shade, especially under trees and shrubs. These shade plants offer protection from insects for the immature fruit of the pepper plants. So the heat of the peppers plays a crucial role in the reproduction of the plant. And other research has shown that capsaicin is protective against fungal attacks to the plants, so spiciness is a defense against another threat as well.

So when it comes to spicy peppers, capsaicin protects not only the human who consumes the plant's fruit but the very plant itself—a good deal for everyone.

Did You Know . . . If you still have doubts that spices have always been highly valued, the case of Alaric I will erase them. He was leader of the Visigoths who laid siege to Rome in 410 CE. His conquest of the city was a key element in the fall of the Roman empire. Alaric's demands of the besieged Romans included 3,000 pounds (1,363 kilograms) of pepper.

Does chewing gum make you smarter?

IT MIGHT SURPRISE YOU to hear that over the last twenty years or more, dozens of scientific studies have addressed that question. Perhaps you'll be even more surprised that the answers produced by those studies are confusing to say the least. The most that can be said about whether chewing gum makes you smarter seems to be "maybe" or "it depends."

Genius.

Most studies have been straightforward. Two groups of people are given tests of memory, learning, attention, problem solving, or alertness. One group chews gum, the other doesn't, and their scores on these various tests are compared.

Here's an example: Elementary school students were read short stories, told to memorize as much of them as they could, and then asked to recite as much as they could still remember after five minutes, then again after twenty-four hours. There was no difference between the gum chewers and the others after twenty-four hours, but when the delay was a mere five minutes, the gum chewers remembered more of the stories. A slightly different version of the experiment in which children were asked to remember short stories after an hour, or two days, showed gum chewing had no beneficial effect after those significantly longer waits. At least one study suggested that chewing gum *worsened* memory.

Researchers have also looked at gum chewing and the ability to focus. In one study, eight- and nine-year-olds chewing gum did no better than non-chewers at maintaining their attention for the first twelve minutes of a sixteen-minute test, but then did better over the last four minutes. Some sort of delayed effect?

A few studies looking at attention also seemed to suggest that there were some benefits to chewing, but others did not.

It's the same picture for planning and working memory—the memory you use trying to remember a cell phone number as you walk across the room to get your phone. Some studies say gum helps; others find no effect. Math, spatial skills, IQ—in every case there are studies for and against.

In the approximately two dozen studies that have been done, this perplexing mix of results has prevented researchers from coming to any definitive conclusions.

But let's say that there might be some very elusive effect that so far hasn't been pinned down. What could it be?

What about the sugar? Could a (small) blast of sugar released while chewing gum boost brainpower, perhaps by stimulating the release of insulin in the brain? That might work, but sadly some studies have shown that sugarless gum has the same effect (or not) as regular gum.

It has been suggested that the act of chewing itself, gum or not, could stimulate the brain by increasing blood flow in our head. One such experiment asked volunteers to push a button with either their left or their right thumb as soon as a left- or right-pointing arrow appeared on the computer screen. Those chewing (odorless and tasteless gum to eliminate those effects) were much faster, and the chewers' brain scans showed higher activity in parts of the brain responsible for both attention and action.

Yet the question about how long-lasting any effect is still lingers. Another study showed better performance for gum chewers on tests like repeating a sequence of numbers backward or solving logic puzzles. (Sugar wasn't important—half the gum chewers were given sugar-free gum.) But the positive effect lasted only twenty minutes, even though the chewing continued.

A scenario that might make sense is that chewing *before* mental testing gears up the brain, possibly by increasing blood flow, but chewing *during* testing doesn't because the brain can do only so much. That would also explain why improvements, if they happen, last only for a while.

Sad that's there not much consistency—it makes it harder to argue for gum chewing during class. As a learning tool, that is.

History Mystery

What are the Nazca lines in Peru?

It's much easier to describe the Nazca lines than to figure out why they exist. They are a vast network of straight and curved strips of bare earth, running across the Peruvian desert, about a six-hour drive south of Lima. Most lines are about a footpath wide (although some are wide enough to accommodate several people walking side by side), but it's the scope and size of the lines that makes them so remarkable.

There are 1,300 kilometers (800 miles) of lines, arranged in several hundred designs, including triangles, zigzags, spirals, and many images of animals: birds, fish, plants, a monkey, and a spider. The range in size is dramatic: The animal shapes are relatively small (100 meters across, or about 330 feet), but some of the bigger geometric figures stretch tens of kilometers. Smaller images are being discovered all the time, sometimes on hillsides just above the desert.

Making them was apparently pretty easy. The high altitude desert in south-western Peru gets little rain or wind, and the people who drew the lines,

The Nazca Lines

perhaps just less than two thousand years ago, simply scraped the superficial stones off the underlying pinkish sand. The color contrast makes the lines stand out. In this sense they are etchings, not nearly as challenging to create as the pyramids or Stonehenge.

Anthony Aveni, a longtime professor of astronomy, archaeology, and native studies at Colgate University in New York, is one of the prime investigators of the Nazca lines. In the 1980s he and a crew of ten volunteers made their own Nazca figure (they're called geoglyphs): a 35-meter-long spiral (almost 115 feet). They had no tools and no plan, and took about an hour and a half to finish it. A more elaborate effort involving twenty people—using broomsticks and string this time—created a long straight line in one working day. To trace all the lines and designs we see today in Peru, Aveni figured that 100,000 pickup trucks worth of rock would have to be swept up. He estimated that a team of ten thousand could have produced every single image on the desert in less than a decade.

So how the Nazca lines were made isn't really an issue. Nonetheless, Aveni contends that the lines are right up there with the Seven Wonders of the World because, like the original seven, they are unique, timeless, and seemingly beyond the capacity of those who built them.

You might puzzle over that last claim because you've just heard that the lines themselves are relatively easy to create. But it's how the enormous Nazca patterns were planned out that is the real achievement.

The lines are best seen from the air. If you were standing on the desert floor next to, say, the monkey pattern, it would be difficult to tell that the line of the monkey's curved back and the one that makes its carefully etched fingers is one single *unbroken* line because the image is about a hundred meters across. Some of the lines of the bigger designs are so large they can be seen from Earth orbit, and so accurate that one triple set runs ten kilometers (more than six miles) with almost no deviation (half of one degree) from perfectly straight. And yet the makers managed all this without helicopters or surveying equipment.

How these lines were designed is not the only mystery. There is also the question of *why* they were made. Complicating the understanding of the lines and the animal and plant figures is the fact that they were created at different times—the animals much earlier than the geometric patterns—raising the possibility that there were completely different reasons for making them.

A German woman named Maria Reiche devoted her career to defending the idea that the lines constituted a giant astronomical observatory. There are indeed some intriguing individual astronomical alignments. One Nazca line points to the setting of the sun on the first day of winter in the Southern Hemisphere, June 21. Reiche also argued that several lines point to what would have been the rising and setting of one of the stars in the Big Dipper at around the time the lines were originally laid down.

But it's a tough argument to make persuasively because there are so many risings and settings of planets and stars that some are bound to line up with the lines, and we have no idea if any of these astrological bodies or dates meant anything to the Nazca designers.

Indeed, Anthony Aveni tested all the lines and found no set of statistically significant associations to astronomy, although he allows that some of the lines might be aligned to dates in November when floodwaters descend from the nearby Andes.

If the lines don't have an astronomical purpose, perhaps they fulfill a more earthbound one. Anthony Aveni discovered that many of the lines are arranged in radial patterns and intersect with other networks of lines. Could the lines simply be paths for walking, and the squares and trapezoids areas for gathering, maybe for religious purposes?

Also, water plays some sort of role. Many of the lines connect with areas where water surfaces, or where there are underground movements of water, but the exact connection between the lines and water is still obscure.

Could they be art with no practical intent? Given that we know so little about the people who made them, it's hard to say yes or no to that.

While most of our questions about the Nazca lines remain unanswered, one theory should definitely be trashed. I am referring here to the ideas of Erich von Däniken, whose enormously popular book about ancient astronauts, *Chariots of the Gods*, earned him nothing but derision from actual researchers, yet racked up huge sales. Von Däniken pounced on the fact that the designs are best seen from the air and claimed they were landing strips for ancient astronaut aliens, who periodically visited Earth to see how we were doing.

Von Däniken even identified the etching of the leg bone of a condor as the place where the aliens turned their spacecraft around on the ground. But it's three meters (ten feet) long—suitable only for an extremely tiny spacecraft, for extremely tiny aliens.

And there is yet another ironic weakness in von Däniken's fabulous conjecture. It's not true that the lines can be seen only from the air. In fact the nearby foothills offer a great view of the amazing designs.

Acknowledgments

You can't write a question and answer book without questions! I thank all those at Simon & Schuster, my friends, and random people online for providing the lion's share of the questions in this book. Of course questions aren't much without answers, and I am indebted both to the incredibly diverse world of science for providing research that answers, or at least partly answers, some of the most esoteric questions. In addition, Christof Reuter; Louise Barrett; George Eleftheriades; Amelia Hunt; my old friends Jearl Walker, Stephen Morris, and Corey Keeble; John Wettlaufer; Young-Hui Chang; Chip Taylor; Mark Changizi; Jayne Yack; Glenn Morris; and Lori Montgomery all provided invaluable insights into the various phenomena in this book.

As had been the case with the previous two *Science of Why* books, my colleagues and friends Niki Wilson and Joanne O'Meara did essential pieces of research and consultation. For instance, Niki dug into the vexing issue of whether sharks can smell a drop of blood a mile away. And wherever there is physics in the book, Joanne helped set the record straight. (She would of course say that it's *all* physics.)

Anyone who writes knows that the more eyes that light on the page, the better the writing will be. It was great to work once again, after many years, with Meg Masters as my editor. She stepped in when Nita Pronovost was assigned greater responsibilities . . . than me. But I'm confident Nita's eye passed over the prose at least once. Brendan May pulled the topics together and Catherine Whiteside is, right now, making the world aware that this book exists.

And, of course, thanks to Kevin Hanson. Once upon a time years ago we talked about doing a book together, but three in a row? And sushi to boot?

I don't need to name the curious—and often skeptical—friends who contribute more than they realize to books like these. There wouldn't be nearly the incentive to write them if they stopped.

Thanks, too, to Mary Anne, who came up with the idea that a "workation" is the best way to write a book! I'm also lucky to have a curious and fun-loving extended family, and we stay in touch no matter where they are. This book is dedicated to the newest member of that group.

Photo: Richard Siemens

Jay Ingram has written sixteen books, including the bestselling first and second books of this series, *The Science of Why* and *The Science of Why²*. He was the host of Discovery Channel Canada's *Daily Planet* from the first episode until June 2011. Before joining Discovery, Ingram hosted CBC Radio's national science show, *Quirks & Quarks*. He has received the Sandford Fleming Medal from the Royal Canadian Institute, the Royal Society of Canada's McNeil Medal for the Public Awareness of Science, and the Michael Smith Award for Science Promotion from the Natural Sciences and Engineering Research Council of Canada. He is a distinguished alumnus of the University of Alberta, has received six honorary doctorates, and is a Member of the Order of Canada. Visit Jay at **JayIngram.ca**.

🐦 @jayingram

Want more curiosities and mysteries?

Make sure to check out *THE SCIENCE OF WHY* and *THE SCIENCE OF WHY²*, the first two volumes of the bestselling series, to discover more fascinating facts and fictions about black holes, electric eels, Bigfoot, and more!

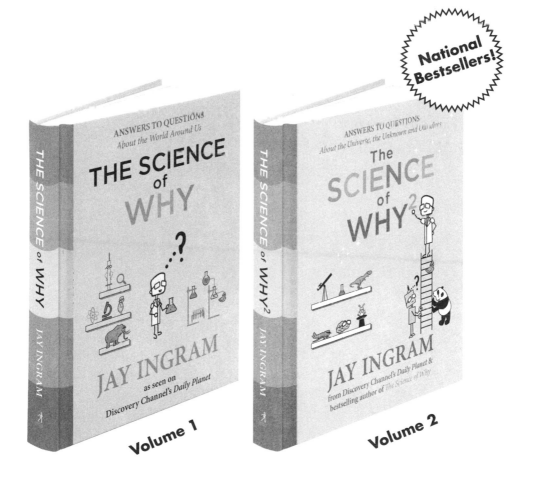